T0134910

Control of Operation Modes of Gas Consumers in the Event of Gas Supply Disruptions

Viktor I. Rabchuk · Sergey M. Senderov ·
Sergey V. Vorobev

Control of Operation Modes of Gas Consumers in the Event of Gas Supply Disruptions

Viktor I. Rabchuk
Melentiev Energy Systems Institute
Irkutsk, Russia

Sergey M. Senderov
Melentiev Energy Systems Institute
Irkutsk, Russia

Sergey V. Vorobev
Melentiev Energy Systems Institute
Irkutsk, Russia

Translated by
Svetlana V. Steklova
Irkutsk, Russia

ISBN 978-3-030-59733-7 ISBN 978-3-030-59731-3 (eBook)
https://doi.org/10.1007/978-3-030-59731-3

This Springer imprint is published by the registered company Springer Nature Switzerland AG
The registered company address is: Gewerbestrasse 11, 6330 Cham, Switzerland

Preface

This research monograph addresses the issues of control of the operation of natural gas consumers in the industrial sector, the electric power industry, and the public utilities sector under conditions of reduced, stopped, and restored gas deliveries. Emphasis is put on the algorithm of the transition of gas consumers from normal to emergency operation and the algorithm of recovering normal operation after an emergency in the gas supply system is terminated.

The authors believe that the book might prove useful for undergraduates majoring in energy-related disciplines, researchers, the personnel involved in the provision of housing and community amenities as well as those working in energy industries and employed by industrial enterprises, where energy consumption mix includes natural gas.

The research was carried out under State Assignment, Project III.17.5.1 (reg. no. AAAA-A17-117030310451-0) of the Fundamental Research of Siberian Branch of the Russian Academy of Sciences.

Irkutsk, Russia

Viktor I. Rabchuk
Sergey M. Senderov
Sergey V. Vorobev

Introduction

In the event of large-scale emergencies in the national gas supply systems (NGS), when the capacity to supply natural gas to its consumers sharply and rapidly decreases, the problem of enforcing hard caps on deliveries to such consumers arises. Moreover, enforcing such caps should be rigorously grounded. This should take into account, first of all, the consumer's requirements with respect to the minimum demanded gas deliveries to them in the first hours after the onset of an emergency so as to prevent the emergency shutdown of the gas consumer with subsequent severe economic and social damage.

The present study covers the control of operation modes of some of the most important gas consumers in the industrial sector, all types of gas consumers in the electric power industry and public utilities sector under conditions of a reduction, cessation, and restoration of gas deliveries in the event of large-scale disruptions in the operation of gas supply systems. Within the scope of the issues covered by the study

- we define the principles of assessing the degree of meeting the demand for uninterrupted gas supply under emergency conditions in gas supply systems for the above gas consumers;
- we divide all considered gas consumers into homogeneous groups (so as to make all gas consumers assigned to a certain group *share* the type of products produced or work performed, operation and process specifics, and the degree of importance of supplying them with gas during an emergency);
- we identify key operation and process specifics of each considered homogeneous group of gas consumers, including the specifics of transferring processes that use natural gas at the facilities of each such group from rated gas consumption to minimum gas consumption with the subsequent fail-safe shutdown;

  – we estimate required gas volumes and the period of time required to resume full-on production by the gas-consuming facilities under consideration (i.e. after recovering their normal operation) upon the termination of an emergency.

As a result, for each considered homogeneous group of gas consumers, the procedure (algorithm) of the rational transition from operating under normal gas supply to operating under emergency gas supply (or the transition to the complete fail-safe shutdown of the gas consumer operation) is presented, along with the algorithm of recovering the normal mode of gas supply for the given gas consumer after an emergency in the gas supply system is terminated. In this situation, both the algorithm of the transition from normal to emergency operation and the algorithm of recovering normal operation boil down to hourly schedules of required volumes of gas supply to the given gas consumer (in the former case, from the onset of the disruption of gas supply until reaching stable emergency mode operation or until the complete cessation of gas supply; in the latter case, i.e., the recovery of normal operation, starting from the termination of an emergency to the complete recovery of such normal operation).

This book was written by the authors affiliated with the Melentiev Energy Systems Institute of the Siberian Branch of the Russian Academy of Sciences (SEI SB RAS).

The research findings presented in this study can be used to substantiate the restrictions on gas supply to administrative-territorial entities and entire regions of countries for the duration of large-scale emergencies in the NSG, when the structure of gas consumers in these entities, the key operation and process specifics of these gas consumers, and the degree of importance of uninterrupted gas supply to gas consumers in case of gas supply disruptions are to be taken into account.

Contents

Chapter 1
Issues Related to the Control of the Operation of Various Gas Consumers in the Event of Their Gas Supply Disruptions

1.1 The Essence of the Study and Its Problem Statement

Under normal operating conditions, the energy system (ES) supplies consumers with the required amount of fuel and energy resources (FER) of the specified parameters. At the same time, the production capacity of a particular ES meets at least the available demand for its products.

In the event of an emergency in the system of fuel and energy supply of the consumer of fuel and energy resources, resulting in the disruption of the operation of this system, the process of the transition of this consumer to a new mode of its operation begins, both in terms of the level of supply of the corresponding type of fuel and energy resources to them and in terms of performance of their functions. For any FER consumer, a new (non-standard) level of its operation can be either a zero level or an intermediate one (that between the initial and zero levels). This new level will be maintained until the emergency is terminated. Afterward, the transition of both the consumers of FER and the systems of fuel and energy supply to this consumer begins so as to recover the previous (normal) level of operation (of course, only if this level was changed with the onset of the emergency). Hence arises the problem of rational control of operation modes of the FER consumer during the above transitions.

For a considerable part of European countries, as well as for the European part of Russia, natural gas serves as the main fuel. For example, in a significant part of these Russian regions the share of gas in the balance of boiler and furnace fuels exceeds 90–95%, and sometimes reaches 99%.

Any developed gas transmission network on the national scale includes an extensive system of gas trunklines combining the sources of gas inflow into it (main compressor stations at gas fields, terminals for receiving imported gas, both pipeline gas and LNG) with points of gas delivery to gas distribution networks, and, in the case of large industrial consumers or export terminals, directly to points of reception of trunkline gas. Any developed gas supply system along with the linear part (trunk pipelines, including compressor stations) includes gas storage systems as well. There

V. I. Rabchuk et al., *Control of Operation Modes of Gas Consumers in the Event of Gas Supply Disruptions*, https://doi.org/10.1007/978-3-030-59731-3_1

are many studies that deal with the issues of increasing the reliability of timely supply of the required volumes of gas through the inclusion of numerous underground gas storage facilities into the gas industry operation [1–4 and others]. At the same time, it is the underground storage facilities of the seasonal type, i.e., designed for seasonal leveling of the pipeline network load profile, that have so far been the most widely adopted.

The complex, often long-established, territorial structure of the national gas supply systems leads to potentially significant problems in ensuring the reliability of its operation. As far as industrialized countries are concerned, it is often the large gas consumers, primarily the powerful industrial centers, that have no gas reserves available in close proximity to them. As a case in point, in [5–7, etc.] the authors studied global natural gas supply chains. Such studies show that due to increased gas deliveries from gas-rich regions to the regions that are short of gas but have energy-intensive industries therein, the global trade imbalance in natural gas is growing.

For instance, even in Russia that is self-sufficient with respect to the natural gas supply, more than 80% of it is produced in a single gas-producing region (Nadym-Pur-Tazovsky district, Tyumen region). This area is 2–2.5 thousand km away from the main gas consumption areas. This situation leads to the fact that (as it is the case in Russia) significant volumes of natural gas are transported over long distances through gas trunkline systems with a large number of mutual crossings and cross-connections, moreover, the runs of powerful gas trunklines are often laid at short distances from each other. Thus, most national gas supply systems have a large number of process facilities critical to the smooth operation of their systems. In the event of a loss or a significant reduction in the performance of such facilities, a significant undersupply of gas to its consumers may occur. Such undersupply can cause significant economic and social damage (depending on the specifics of the consumer) with a whole array of negative consequences in the relevant areas of the economy and life in general. This situation is typical for both the Russian and the European gas pipeline networks, the latter being national gas transmission networks technologically linked as a unified system [8–11].

Speaking about the national gas supply systems of the majority of geographically distributed countries as a whole, one should not forget about specific regions of that country and specific consumers and their types. The natural gas needs of each specific region and each gas consumer in a given region must be met to the maximum extent possible in the event of any emergency in the gas industry. Naturally, a stable supply of natural gas and a reduction in interruptions of its distribution to consumers support the sustainable development of countries and regions. This has been the subject of a good deal of published research. So, for example, in [12] security of natural gas supply is considered as a hierarchical and multidimensional structure. It is shown that gas supply security in the context of mitigating gas supply disruptions to consumers includes three main aspects: availability, infrastructure, and control. Availability can be achieved not only by adequate supply, but also by efficient energy use. Besides that, behavioral and technological changes are also important. When we speak of the need to ensure reliable supplies of natural gas to consumers, it is natural to consider not only pipeline gas supplies but also LNG supplies. For example, in [13] this issue is treated as based on the case study of Iran. To this end, the methods of gas delivery to consumers in the form of both compressed natural gas and LNG are analyzed. In [14], it is also stated that the current dependence on the oil and gas industry for the purposes of economic development and social activities calls for a study of the sustainability of supply chains of respective energy resources. In [15], in addition to the established methods of reliability analysis, a new methodology of gas supply reliability assessment of gas transport systems is proposed. Taking into account the uncertainty of both gas supply and demand, the calculations of these two elements are combined into a single Monte Carlo model. In each Monte Carlo trial, the hydraulic analysis of the non-stationary flow is combined with the simulation of the state transition process to calculate the gas supply capacity. In addition, the study investigates the impact of uncertainty in supply volumes and market demand on the reliability of gas supplies and proposes certain measures to improve gas supply reliability.

In connection with the above, the issue of managing the operation modes of various gas consumers in the event of disruptions of natural gas supplies to them, including the issues of substantiation of rational profiles of hourly supply of this gas to specific types of its consumers during the transition of the gas consumer from the normal mode of gas supply to the non-standard mode and the reverse transition back to the normal mode, becomes relevant.

Each gas consumer, depending on the specifics of gas consumption and the type of their activity, has its own individual characteristics. Therefore, to consider the above issues individually for each gas consumer is deemed an obviously unfeasible task. In any case, they (i.e., gas consumers) should be treated as groups. A specific group of this kind could be a set of gas consumers producing one specific type of product or performing one and the same specific type of work. In Table 1.1 such groups are shown as applicable to the majority of industrially developed European countries as exemplified by Russia. In this table, the specifics of natural gas consumption in each such group is indicated by the relative volumes of natural gas use in the total fuel consumption for the production of respective goods and services. The table also provides details on the share of each group in the total natural gas consumption in Russia.

Gas consumers of most of the groups listed in Table 1.1 are linked not only by the commonality of the type of products (or the type of work performed) but also by the commonality of the operation and process specifics and the importance of providing them with gas during emergencies (for example, the production of cast iron in blast furnaces, open-hearth steel production, cement production using gas, etc.). Such groups (bringing together gas consumers not only based on the type of products produced or work performed but also based on their operation and process specifics, as well as the importance of providing them with natural gas) *can be considered homogeneous*, with the issues discussed above to be considered as targeting them.

However, Table 1.1 also contains such groups, within which different gas consumers, while producing the same product, use fundamentally different processes,

Table 1.1 Share of natural gas in total fuel consumption by main groups of gas consumers in Russia (by type of goods produced and services provided)

Gas consumer groups by type of goods produced and services provided	Share of natural gas in total fuel consumption	Group share in total natural gas consumption
	%	
Electricity generated by power plants running on boiler and furnace fuel	65–68	30–35
Heat generated by power plants running on boiler and furnace fuel	70–73	16–17
Heat generated by boilers	72–75	17–19
Generation of electricity and heat, total	65–68	65–70
Iron and manganese ore sinter	5	<0.1
Iron ore pellets	75–80	0.1
Metalized iron ore pellets	100	0.2
Blast furnace air	36–37	0.2
Compressed air	45–48	<0.1
Heating of hot-blast stoves	3	<0.1
Cast iron	20–21	1.3
Open-hearth steel	65–68	0.2
Rolled ferrous metal products	70–75	0.8
Steel pipes	100	0.2
Cast iron pressure pipes	100	<0.1
Heating of coke oven batteries	1	<0.1
Synthetic ammonia	100	<0.1
Iron castings	40–45	<0.1
Steel castings	80–85	<0.1
Heat treatment of metals	83–88	0.2
Cement	75–80	<0.1
Clinker	90–95	1.6
Structural glass	100	<0.1
Direct use of gas by households	100	11–12
Other consumer groups	42–45	15–20
Total consumption for the production of specified types of products or work in Russia as a whole	42–45	100

have different operation and process specifics, and a different degree of importance of ensuring gas supply during the emergency. Such groups include gas consumers producing electricity and heat, i.e., gas consumers in the power and public utilities sectors (the latter sector also includes the direct use of natural gas by households).

Two aggregated groups (the power and public utilities sectors) consume almost 80% of the gas used domestically. In order to arrive at homogeneous groups, each of the above aggregated groups (both in the power and utilities sectors) is divided into smaller groups, within which each gas consumer has approximately the same operation and process specifics and the same degree of importance of uninterrupted gas supply during an emergency. Only such division allows to arrive at homogeneous groups in these gas consumption areas. The principles behind such division of the gas consumers under consideration into homogeneous groups are given below, at the end of this Sect. 1.1.

It is fair to assume that for each homogeneous group of gas consumers (related by the common type of products or the type of work performed, by the common operation and process specifics and the commonality in relation to the degree of importance of uninterrupted gas supply to these consumers during an emergency under a given scenario), it is possible to find an optimal (rational) schedule of the transition from the normal (regular) operation of consumers and the normal level of gas supply to new levels (for the duration of the emergency). Such rational schedules, apparently, can also be defined in the case of restoration of normal gas supply modes of consumers after the emergency. In order to get to the above schedules of the given homogeneous group of gas consumers, it is necessary to know the main characteristics of the required change in the modes of operation of these gas consumers under conditions of a reduction, termination, and restoration of gas supplies to them in the event of gas supply disruptions. Obtaining these characteristics requires taking into account the peculiarities of gas consumers of this homogeneous group in terms of possibilities and duration of their full or partial switchover during the emergency to other types of fuel and energy resources and reverse switchover back to gas after the the emergency is terminated, as well as such operation and process specifics of these gas consumers as:

- characteristic features of those elements (facilities) at the gas consumer, the operation of which is noticeably disrupted when gas supply is reduced or ceased completely;
- the nature of dependencies reflecting the level of decline in the output of its products by this group of consumers at different levels of undersupply of gas to the consumer;
- maximum depth of the gas supply reduction that, when exceeded, makes gas consumers of this group stop producing their products (or providing respective services);
- special aspects of transffering technological processes from the normal mode of gas consumption to the mode of reduced gas consumption up to subsequent fail-safe shutdown;

- details on the required hourly volumes of gas supply to the consumer at different points in time when the gas consumer restarts production after the emergency is terminated, as well as details on the total time required for the gas consumer to return to the previous level of production (the one available before the emergency).

Thus, *the control of modes of operation* of different kinds of gas consumers under conditions of decrease, cessation, and restoration of gas supplies to them *boils down to substantiation of rational schedules of the transition* of considered homogeneous groups of *gas consumers from the normal mode of their gas supply to a non-standard mode* up to a complete fail-safe shutdown of gas consumers and to *substantiation of rational schedules of restoration of the normal mode* for the said homogeneous groups. In its turn, such substantiation itself is impossible without all the details listed above, as well as the data on the importance of uninterrupted gas supply to various gas consumers during an emergency.

With respect to the importance of uninterrupted gas supply during emergencies in the gas supply system, in our case all gas consumers are divided into four categories.

The first category is gas consumers whose gas supply system disruption may lead to an unacceptable weakening of the country's defense capability.

The second category is made up of consumers whose gas supply disruptions may lead to fatalities, significant health deterioration, and a sharp deterioration in the quality of life of large groups of population, whose life support is related to the operation of these gas consumers.

The third category is gas consumers, the failure of gas supply to whom can lead to significant economic losses, including the destruction or deterioration of their main production assets, to a sharp imbalance of production and process relationships between producers and consumers of the relevant products, to a significant deterioration in product quality, etc.

The fourth category is the gas consumers the disruption of gas supply to whom will not lead to those consequences that are specified for consumers of the first, second, and third categories, i.e., to gas consumers of the given (fourth) category belong all those that are omitted from the first three categories.

The book covers three classes of gas-consuming facilities: industrial, power, and public utilities facilities. Each class is made up of its own gas consumers that differ significantly from each other in the degree of importance of gas supply during an emergency.

At this point, it should be noticed that the authors do not single out and do not consider those gas-consuming facilities that can be relegated to facilities of the first category (facilities whose operation in no case should be disrupted because of failures of gas supply systems), at least for the reason that when such facilities are designed and constructed their system of gas supply in case of the failure of its operation should be necessarily backed up by other system of power supply (for example, independent power supply from the diesel power plant with stocks of diesel fuel available).

Next, it is necessary to determine whether the gas consumers under consideration in the industry, power sector, and public utilities sector belong to one of the three remaining categories (second, third, or fourth). Based on the substantive nature of these categories, it is possible to determine the specified *membership* also only as a result of the analysis: on the one hand, that of *the operation and process specifics of gas consumers of this homogeneous group, and, on the other hand, that of the economic and social significance of gas consumers of the same group.* The problem of dividing gas consumers into homogeneous groups arises again when it is necessary to know the features and importance of each individual gas consumer while accepting the practical impossibility of obtaining such details, if the number of different gas consumers under consideration, as is in our case, will be quite large.

To address this issue, it was deemed feasible to follow the below path. Among industrial gas consumers (whose diversity is too great), the most technologically complex and the most important for the economy ones have been selected. These are ironworks and non-ferrous smelters, the most significant sub-sector of the construction materials production sector, as well as one of the types of consumers where gas is used not only as an energy carrier but also as a raw material.

The consumption of natural gas by the electric power industry and public utilities sector (including direct use of gas by households), as mentioned above, accounts for almost 80% of all gas consumed in the country. This, as well as the wide variety of processes used to produce electricity and heat with natural gas, requires a very careful approach to the division of such gas consumers into homogeneous groups.

Gas consumers in the electric power and public utilities sectors are divided *into homogeneous groups on the basis of the following information* on these gas consumers with respect to the key operation and process specifics and their importance for the economy and in social terms:

- the type of products produced or services provided by this group of gas consumers (single or dual product production;
- a possibility to transfer gas consumers of this group from the rated gas supply mode to the minimum supply and subsequently fail-safe shutdown mode;
- a possibility of the switchover for gas consumers from one type of FER to another (from gas to another type of fuel and back to gas after the emergency is terminated);
- the degree of importance of uninterrupted gas supply to the consumers of this group in the broad sense of the very concept of the degree of importance (i.e., taking into account such factors as the possibility of fatalities in case of termination of gas supplies, a large economic loss from the termination of production by the gas consumer due to gas supply disruptions, a drastic deterioration in the quality of life of the population related to the operation of this gas consumer, etc.).

At the same time, the purely qualitative plan should take into account such information as the dependence of the output of their products by this gas consumer on the level of undersupply of gas to them, as well as the depth of a reduction in gas supplies to the gas consumer that triggers their stopping the production of their products.

All of the above characteristics were used to develop specific criteria for division into homogeneous groups of gas consumers in the electric power and public utilities sectors. These criteria themselves and the results of dividing gas consumers into homogeneous groups based on such criteria are presented in Sects. 3.1 and 4.1.

A comprehensive analysis of obtained homogeneous groups of gas consumers can change the researcher's understanding of the importance of uninterrupted gas supply to a group of gas consumers or to individual gas consumers in a group. In this case, an appropriate adjustment should be made for the latter (individual groups or individual gas consumers).

1.2 The Procedure for the Study of Main Patterns of Changes in the State of Gas Consumers in the Event of Gas Supply Disruptions

For all the considered homogeneous groups of gas consumers (Section 2: industrial gas consumers, Section 3: gas consumers in the electric power industry; Section 4: gas consumers in the public utilities sector) it is expedient to adhere to the following general procedure for studying the above patterns to arrive eventually at specific measures to change their mode of operation due to gas supply disruption.

- Identification of specific operation and process specifics of gas consumers of a given homogeneous group.
- Assessment of the degree of importance of uninterrupted gas supply to gas consumers of a given group, taking into account the operation and process specifics of these gas consumers and factoring in the social and economic importance of gas consumers of a given group to the country.
- Getting specific measures related to the change in the mode of operation of gas consumers of a given homogeneous group (including the need for a complete fail-safe shutdown) in connection with a reduction in gas supplies (or termination thereof) and getting to the same measures, but related to the restoration of the regular gas supply mode. These specific measures include substantiation of the hourly schedule of the minimum required volumes of gas supply to the specified consumers in case of its supply reduction (so as to continue the operation of gas consumers under relatively normal conditions and during the emergency period in their gas supply systems), as well as substantiation of the rational hourly schedule of gas supply to its consumers in case of restoring the normal mode of operation after the emergency is terminated.

The meaning and required content of each of the above steps are discussed below.

Identification of specific operation and process specifics of gas consumers of a given homogeneous group. First of all, such features include *the list of those elements of the process flow of gas consumers* of a given group (facilities and equipment), *the operation of which is disrupted when the gas supply to them is reduced or ceases*. Here one needs to perform a thorough analysis of the significance of each such element for normal operation of a given group of gas consumers under consideration and in the event of disruptions of gas supply mode of this element.

The results of such an analysis allow to proceed to consideration of the following operation and process specifics of gas consumers of a given homogeneous group: *the features of the transfer of such gas consumers from the rated mode of gas supply to the minimum one*, up to complete failsafe shutdown of the gas consumer (if the degree of disruption of gas supply will require this shutdown). The study of the above feature will allow forming an hourly schedule of supply of the minimum necessary volumes of gas for the main gas-consuming elements of a given group of gas consumers from the moment of the onset of an emergency in gas supply systems up to the moment of arriving at a stable (though non-standard, i.e., reduced) mode of gas supply or up to stop the operation of a given group of gas consumers completely. Such an hourly schedule does not take into account the possibility and duration of gas consumers' switchover (complete or partial) from gas to other fuels. It is advisable in this case to present hourly schedules of the minimum required gas supply to gas consumers of a given homogeneous group as a percentage of required gas supply volumes under normal regular gas supply mode, because gas consumers of this group, sharing their operation and process specifics, can be of different scale in terms of capacity (output of their products), and hence differ in terms of cardinal values of the volumes of gas consumed.

Possibilities and duration of the switchover of gas consumers of a given homogeneous group from gas to another type of fuel is the next operation and process feature that needs to be addressed. The result of such consideration should be as follows:

- the list of gas-consuming elements of this homogeneous group of gas consumers, where gas can be substituted for other fuel during an emergency;
- maximum possible degree of gas substitution for other (specific) type of fuel for each gas-consuming elements in a given group of gas consumers;
- the required duration of switching over the gas-consuming element from gas to such fuel;
- the adjusted hourly schedule of supply of the minimum required volume of gas to gas consumers of a given homogeneous group (for the time being, taking into account only the possibilities and duration of the switchover of the gas consumer from gas to other fuel) from the moment of the beginning of gas supply disruption to getting to the stable mode of gas supply (reduced, non-standard) or to the mode of complete termination of gas supply (and consequently to the stable mode of supply, partial or full, by the substituting fuel).

Next, one shall consider *such operation and process specifics* (for a given homogeneous group of gas consumers) as the dependence of the output of its products on the level of undersupply of gas, taking into account the substitution of gas for other fuel, as well as the depth of a gas supply reduction, at which gas consumers stop production. The results of the study of these features will serve as a prerequisite (along with other information) for assessing the degree of importance of uninterrupted gas supply to various gas consumers during emergencies in gas supply systems.

The next stage of the study of the operation and process specifics is the review of *the specifics of gas consumers of a given homogeneous group with regard to the restoration of their normal mode of operation after the emergency it terminated.*

Here, such specifics, firstly, are related to the required duration of the switchover to gas from other fuel (if during an emergency there was a corresponding switchover by the gas consumer from gas to other fuel), and secondly, these are the required volumes of gas and the required time for the restoring of operation and production of its products by the gas consumer of a given group. This concludes the study of the operation and process specifics of a given homogeneous group and in accordance with the adopted general procedure of the studies one makes an assessment of the importance of uninterrupted supply of gas to the indicated gas consumers.

Assessment of the degree of importance of uninterrupted gas supply to consumers during the emergency should take into account their operation and process specifics (peculiarities of the transition of the gas consumer from the rated mode to the minimum one, the possibility and duration of the switchover by the gas consumer during the emergency from gas to other fuel, the dependence of the output of its products by this gas consumer on the level of undersupply of gas, the required volume of gas and the required time to resume production of this gas consumer in full). But first of all, such an assessment should take into account the social and economic importance of gas consumers of a given homogeneous group (the possibility of damage or destruction of gas consumer's basic production assets in the event of termination of gas supplies, reducing the quantity and quality of products of the gas consumer, the possibility of a sharp deterioration in living conditions of people associated with the disruption of the operation of a given gas consumer, etc.). The results of the above assessment are taken into account when arriving at the rational schedule (algorithm) of the transition of gas consumers of a given homogeneous group from the standard mode of their gas supply to a non-standard mode, as well as when arriving at the rational schedule (algorithm) of restoring the standard mode.

Chapter 2
Control of Operation Modes of Gas Consumers in the Industrial Sector in the Event of Their Gas Supply Disruptions

As mentioned above, an exhaustive review of all gas consumers in the industrial sector is not to be undertaken because of too wide a variety of such gas consumers with respect to the type of products manufactured (or the type of work performed), operation, and process specifics, and the importance of their uninterrupted gas supply during emergencies in gas supply systems. Herein, the scope of consideration was limited by the following:

– the fully integrated steel mill: from facilities for the heat treatment of iron ore raw materials to heating furnaces in rolling, including also an industrial buffer storage CHP plant located at the mill premises, and repair shops;
– facilities: gas consumers of the non-ferrous smelting industry, such as aluminum, copper, lead, zinc, and nickel smelting, titanium and magnesium smelting industrial enterprises;
– cement production, as the most gas-intensive industry in the construction materials industry;
– methanol and ammonia production (where natural gas is used as a feedstock).

In carrying out these studies (Chap. 2), we relied on the published research [16–25].

V. I. Rabchuk et al., *Control of Operation Modes of Gas Consumers in the Event of Gas Supply Disruptions*, https://doi.org/10.1007/978-3-030-59731-3_2

2.1 Iron and Steel Industry

2.1.1 Main Elements (Facilities) of the Iron and Steel Industry that Consume Natural Gas and Specific Requirements for Their Gas Supply During Emergencies

For a fully integrated steel mill, all elements (facilities) consuming natural gas are divided into those that are part of the main production and those that are part of the auxiliary production. The volumes of consumption of this gas at such a mill by various facilities of *the main production* are as follows (with a breakdown into years for various similar enterprises in relation to the total consumption of gas by the enterprise):

- facilities for thermal processing of iron ore raw materials—2–3%;
- blast furnace process—4–5%;
- heating of air heaters (hot-blast stoves)—10–12%;
- heating of coke oven batteries—22–25%;
- open-hearth furnaces—30–35%;
- basic-oxygen/oxygen converter process—2–3%;
- basic process heating furnaces—8–10%.

The same volumes of natural gas consumed at auxiliary production facilities are as follows: repair shops—4–5% and buffer storage CHP plants—6–8%.

Almost all of the above facilities (both primary and auxiliary) consume natural gas together with coke oven gas and blast furnace gas. The composition of the gas mixture used is determined by process requirements for calorific content of such a mixture for a given gas consumer and a given moment of time (we remind the reader that caloric content of natural gas is 8000–8500 kcal/nm^3, that of coke oven gas is 4000–4500 kcal/nm^3, and that of blast furnace gas is 900–1000 kcal/nm^3).

Facilities for thermal processing of iron ore raw materials. For purposes of preparation of metallurgical raw materials, one employs sintering and firing conveyor machines; so, shaft kilns serve for oxidizing and reduction firing of lump materials and pellets metallization, rotary kilns are for this oxidizing and reduction firing process, along with fluidized bed kilns and combined plants (consisting of the above-mentioned devices).

The main method of preparation of raw material agglomerates for blast-furnace smelting is agglomeration. A layer of furnace charge is loaded onto the moving fire grate, on the pre-stacked safety bed. The upper layer of the furnace charge is ignited using gas burners installed in a special purpose iron receiver. Hot products of combustion (1200–1400 °C) of gas fuel transfer heat to the upper layer of the furnace charge (thus creating prerequisites for the beginning of combustion). Air inflow through the layer is achieved by means of a vacuum chamber. The agglomeration process is quite cost-efficient as it ensures almost full utilization of waste gas heat.

Shaft-type furnaces are used at blast furnace process facilities, during oxidizing roasting of pellets, lime, dolomite, as well as reduction roasting of lump materials (sponge iron, iron ore pellets). Shaft-type furnace designs are determined by the peculiarities of the process, characteristics of auxiliary units for blast supply, and values of required parameters of heat transfer media and reducing gases.

Anecdotal evidence suggests that the greatest economic efficiency in the heat treatment of iron ore raw materials is achieved through the use of a mixture of gases that consist of natural gas (30%), blast furnace gas (60%), and coke oven gas (10%).

As long as there is a stock of readily available sinter (in case of an acute shortage of natural gas) the operation of iron ore heat treatment facilities can be stopped without any consequences for the steel mill.

Coke oven batteries. Coke oven batteries are used to produce coke from sintered (coking) coal by air-free heat treatment. The coke consists almost entirely of carbon (about 90%), has high strength and gas permeability. coke oven batteries are protected consumers of gas fuel and allow no disruption of their gas supply mode. The required calorific content of the heating gas is achieved by using the appropriate proportions of blast furnace gas, coke, and natural gas. On average, out of 100% of the gas mixture volume here 60% is coke gas, 15% is blast furnace gas, and 25% is natural gas.

Coke oven batteries are one of the largest consumers of gas fuel at the steel mill. Furthermore, the process of coal coking in the battery chamber is periodic and is due to the need to turn it off for discharging the finished coke and loading a new portion of the furnace charge. The period of coal coking in the chamber lasts about 16 h. When a single chamber (a separate coke oven battery chamber) is considered, the coal coking process is *a cyclic process*. However, each coke oven battery contains several dozens of chambers so that, in general, the coking process can be assumed to be *continuous*.

Disruption of the operation mode of coke oven batteries with respect to the heat of combustion of the gas mixture used leads to a breakdown of the operation of these batteries. Therefore, when baking another charge of coke in an individual chamber ($\approx$16 h), one can not change thermal conditions for this chamber (if the onset of gas supply disruption in the region where the steel mill is located occurs at this very time). During the transition from the normal to non-standard gas supply mode and during the non-standard (but already stabilized) gas supply mode when baking the next charge of coke in the same individual chamber during the emergency it is possible to reduce or completely stop the supply of natural gas by means of the corresponding increase of the share of coke gas in the mixture of the gas used for heating of coke oven batteries. With that being said, at the same time one will have to accept a certain increase in the "baking" time of this charge (due to the reduced heat of combustion of the gas mixture used).

Thus, speaking of the entire coke oven battery, after the onset of the emergency it is necessary to arrange a gradual chamber-by-chamber substitution of the used fuel mixture that contains natural gas for a mixture with its reduced or zero content.

Blast furnace process. The most difficult process in the ferrous metallurgy is to produce cast iron in blast furnaces. In modern furnaces that operate without interruption for 5–6 years, to intensify the process of iron smelting and reduce coke consumption oxygen and natural gas are supplied along with the high-temperature air blast through the tuyeres. To this end, coke is partially replaced by natural gas, which changes the degree of direct and indirect reduction of iron oxides and increases the combustion temperature in the iron receiver.

Regenerative air heaters (hot-blast stoves) in the blast furnace process are individual gas consumers with their specific features. They operate in a cyclic mode and are mainly heated by blast furnace gas. The blast furnace gas is enriched with natural gas (or coke gas) to increase the blast furnace dome temperature. Air heaters are designed to heat the air blast (up to 1200 °C) before feeding it into the blast furnace. The consumption of combustible gas mixture by air heaters is relatively constant. The quality of the blast furnace process depends on the efficient operation of air heaters, therefore, blast furnace air heaters are among the main protected gas consumers. However, in the case of acute shortages of natural gas, its share in the blast furnace gas mix can be reduced with a barely noticeable deterioration in the blast furnace operation mode.

The approximate consumption mix of gaseous fuel by the blast furnace process is as follows: 90%—blast furnace gas and 10%—natural gas.

Natural gas in the blast furnace is the process feedstock. When stopping the supply of natural gas it is necessary to increase the coke consumption beforehand, i.e., it is possible to stop the supply of natural gas to the blast furnace without disturbing its operation only in 3–4 h after increasing the coke supply.

Steelmaking facilities. Melting units differ not only in purpose and process of steel production but also in methods (stages) of their production and processing, as well as methods of supply of energy carriers and intensifiers. A common method of mass production of steels is the open-hearth method. Open-hearth furnaces have various heating systems based on the use of natural gas, fuel oil, or a mixture of gas and fuel oil. In the last two to three decades, mainly gas-fired open-hearth furnaces have been used. Oxygen and compressor air supplied to the flare and molten bath are used as intensifiers in open-hearth furnaces. Smelting processes are intensified with the help of gas-oxygen burners or gas-and-oxygen tuyeres when installed through a vault or when directly immersed in a molten metal pot.

The steel production process in the open-hearth furnace is cyclical, lasting up to 12 h. The smelting time depends on the steel grade and breaks down into several steps: cast iron pouring, melting, blocking, refining, and steel discharging. Existing open-hearth furnaces are relatively large consumers of natural gas, which is almost the only fuel available to them. Gas supply interruptions cause the furnace bath and molten metal to cool down, which means that the furnace is eventually destroyed. Hence, open-hearth furnaces are the main protected consumers of natural gas at the fully integrated steel mill.

In the basic-oxygen converter production of steel, natural gas is needed for preheating and melting scrap metal or for heating liquid cast iron with subsequent supply of the melt to the converter for blowing oxygen through this melt (carbon burning). Natural gas is also used in electrosmelting, that is for preheating scrap metal in special purpose containers ("buckets"). Drying and heating of lined buckets and containers are also done with natural gas.

Steelmaking facilities that use natural gas *do not allow its substitution with other fuel* and require gas supply to the facility *throughout the entire melting period* (up to 12 h).

Rolling and heat treatment of metals. The metal is heated in furnaces of various designs prior to entering rolling mills and during heat treatment. These furnaces are heated with natural gas, coke oven, blast-furnace gases, or mixtures thereof. Common requirements for all heating furnaces are uniform heating and minimal metal loss due to tarnishing.

Soaking pits are the main type of devices used for heating for rolling on breakdown mills (blooming mills, slabbing mills) of huge metal ingots that have large weight. For heating slabs and blooms, as well as compressed billets of various shapes (circle, plate, square), continuous furnaces having various designs and equipped with various heating systems are widely used. According to the principle of their operation, continuous furnaces are units of continuous operation with characteristic counterflow of metal and combustion products and with zone fuel supply.

Chamber furnaces are used for high-temperature heating of metal (heating of flat hot-rolled bar, heating of breakdown bars, heating of ingots and billets for forging and rolling). The predominant type of fuel for these stoves is natural gas. Sometimes it is used being mixed with blast furnace gas.

On average, it can be assumed that in rolling, 100% of the volume of gas mixture used is 40% blast furnace gas, 40% coke oven gas, and 20% natural gas.

The interruption of gas supply to the heating furnaces during heating of some metal billets or products (bars, rolled products, strips, wires) does not lead to the destruction of the facility itself or any of its elements. In this case, one simply rejects (based on the quality of the metal) the batch of these blanks or products that were heated.

Auxiliary and buffering processes. Auxiliary consumers of gaseous fuel in the steelmaking process include repair shops (cast house, forge shop, heat-treatment shop, etc.), kilns, furnace charge defrosting sheds, CHP plants, and heating boiler houses. Auxiliary consumers of gaseous fuels are usually able to switchover from one type of fuel to another. The role of buffer gas consumers (blast furnace gas, coke oven gas, and natural gas) is relegated to CHP plant boiler units. If there is a need for a sharp increase in gas supply to the main steelmaking process, CHP plant boiler units quickly switchover to other fuel.

2.1.2 Special Aspects of Transferring Ferrous Metallurgy Gas Consumers from the Rated Gas Supply Mode to the Minimum Supply Mode and Their Subsequent Failsafe Shutdown

The total consumption of natural gas for coke oven battery heating accounts for about 60–70% of its total amount supplied to the steel mill in steelmaking and blast furnace process. Interruptions in the supply of the specified amount of gas lead to severe breakdowns and disruptions of operation of the entire plant.

Sinter plants, rolling mill furnaces, repair shops, and buffer storage CHP plants can accommodate restrictions on natural gas supply (up to complete cessation of its supply to the specified facilities) during emergencies in the system of natural gas supply to consumers through the use of a backup fuel. In addition, the listed facilities (except for the CHP plant) can for some time (up to 3–5 days) cease to operate completely (of course, if there is an appropriate stock of, for example, sinter). In this case, the CHP plant can switch over to a backup fuel (fuel oil with an addition of blast furnace gas or coke gas), or fuel oil only, if a reduction in the natural gas supply to the plant will lead to a reduction in the supply of coke gas and blast furnace gas to the CHP plant.

The volume of natural gas supplied to sinter plants, rolling mill furnaces, repair shops, and the buffer storage CHP plant can account for up to 40% of the total volume of gas used in the steel mill. The complete shutdown of the sinter plant and repair shops (or their switching over to a backup fuel), as well as the switching over of the CHP plant to a backup fuel takes approximately 2 h.

The undersupply of natural gas *to coke oven batteries* causes *significant economic damage* due to a drastic change in the high-temperature heating of the furnace batteries. Breaking the furnace heating thermal conditions always leads to the discharge of coke that does not conform with the requirements of the standard and ultimately to the disruption of the entire steel mill cycle. As it has already been mentioned, the very process of "baking" another charge of coke in an individual chamber does not allow to disrupt thermal conditions of heating during the "baking" period. If one simply turns off the natural gas supply exactly during the specified period of time, it will lead to a dramatic disruption of thermal conditions of the battery and discharging poor quality products. This fact does not lead to the deterioration of expensive structures, but to the disruption of the steel mill cycle and to significant economic losses. If one combustible mixture (with natural gas) is substituted for another (without natural gas or with lesser content thereof), naturally, relatively minor economic losses will occur, but they will not be all too noticeable.

Figure 2.1 shows the required hourly supply of natural gas to coke oven batteries of the steel mill during an emergency with gradual chamber-by-chamber substitution of one fuel mixture for another. The duration of such substitution will be determined by the duration of "baking" a charge of coke in an individual chamber (approximately 16 h).

Here (Fig. 2.1), Q is the required hourly level of natural gas supply to coke production (as a percentage of the overall level of supply of this gas to the steel mill under the regular mode of gas supply).

Figure 2.2 shows the required hourly levels and duration of natural gas supply to the blast furnace process at its failsafe transition from the normal mode of operation to a non-standard one ($t = 0$ is the onset of the emergency).

The failure to supply natural gas to the blast furnace process without substituting it for coke may lead to a sharp deterioration of the blast furnace operation and a noticeable economic loss. During this substitution, blast furnaces should be supplied

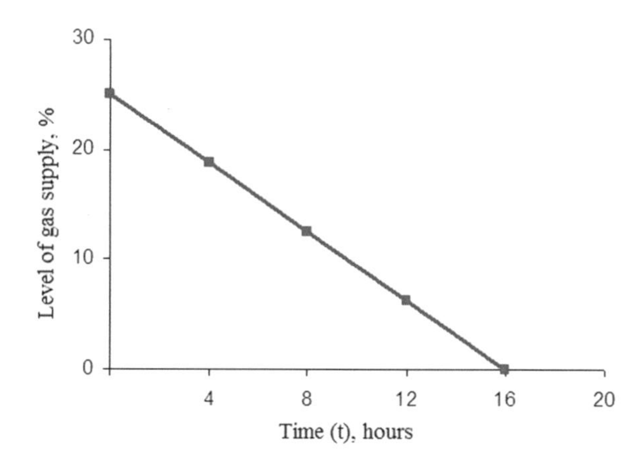

Fig. 2.1 Required hourly levels and required duration of natural gas supply for coke production during its fail-safe transition from normal to non-standard operation ($t = 0$ is the moment of the onset of an emergency)

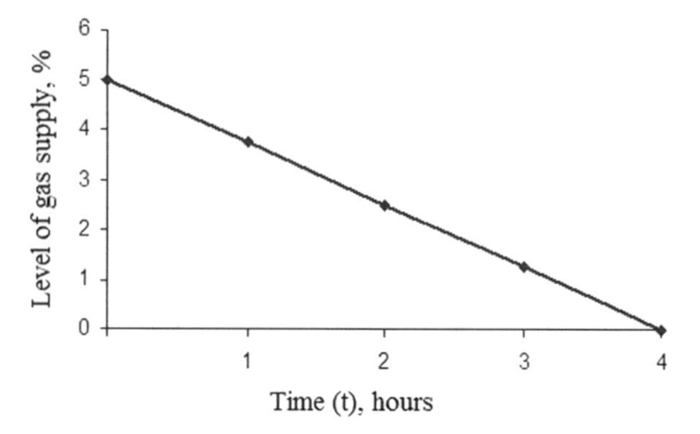

Fig. 2.2 Required levels and duration of natural gas supply to the blast furnace shop as it transitions from the normal mode to a non-standard mode during an emergency

with natural gas in the amount of 5% of the total gas supply to the steel mill when it is operating in the regular mode with respect to gas supply conditions. While natural gas is being substituted for an additional amount of coke (takes 4 h), the supply of this gas to the stoves is stopped.

Figure 2.3 shows the required hourly levels of natural gas supply with the breakdown into melting periods of steel for *an individual* open-hearth furnace under its normal operation mode. During the initial melting period (approximately 4 h), the maximum gas quantity (100%) is fed into the furnace to heat the charge and melt the metal. The amount of gas being supplied is then reduced and by the end of the melting, that is after about 12 h, it amounts to about 50% of the gas supplied to the furnace at the start of the melting.

The open-hearth furnaces fired by natural gas have no backup fuel due to process restrictions. The open-hearth process in the event of gas supply disruptions can be stopped after the end of the currently performed melting without much damage, but only in one case, that is if this process is outside of the integrated steelmaking cycle. The open-hearth shop of the fully integrated steel mill together with the blast furnace shop, coke production, and sinter plant are the most important elements of

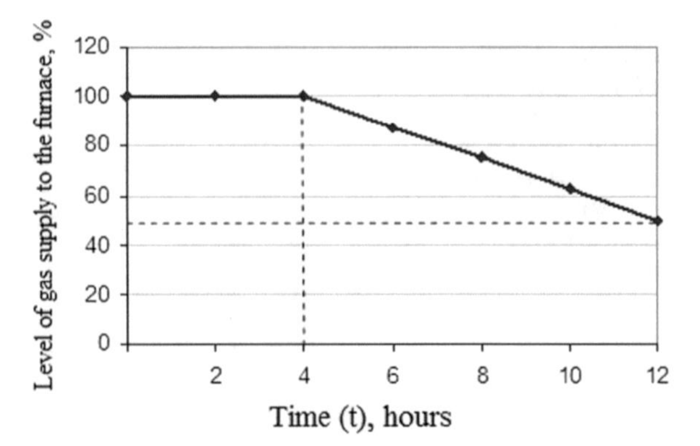

Fig. 2.3 Required hourly levels of natural gas supply to an individual open-hearth furnace given its standard mode of operation (here, Q_m—the level of gas supply to the furnace, % of the level at the beginning of smelting)

the above mill. Shutting down of one of these elements (except for the sinter plant, when there is an adequate stock of sinter available) will make the whole mill cease operating with huge devastating consequences for the mill itself and with noticeable consequences for the economy of the country as a whole. Since the blast-furnace shop, coke production, and sinter plant, as mentioned above, can operate with no natural gas (albeit with some deterioration in performance), while the open-hearth shop can not do away with the gas, in order to avoid large economic damage (or prevent the mill from stopping), natural gas has to be supplied to the open-hearth shop during an emergency in the same volumes as under normal conditions.

Figure 2.3 (slightly above) shows the required hourly levels of natural gas supply to an individual furnace. If there are several furnaces and if, at the moment, each furnace is at its own stage of melting, the required hourly levels of gas supply to the open-hearth shop can be assumed to be constant being equal to 30–35% of the total level of natural gas supply to the mill under normal conditions. These 30–35% should be considered as the minimum required level of gas supply to the mill during an emergency to prevent significant economic damage due to an emergency in the gas supply system.

As a result, the following can be stated with respect to the *possibilities* and *duration* of the transition of the integrated steel mill from the normal mode to a non-standard one without any noticeable economic damage:

- The total duration of the complete transition of the mill from the normal to a non-standard mode of operation is determined in this case by the duration of bringing the charge of coke to the ready state in the chamber where the coking process has just started at the moment of the emergency onset, i.e., the duration of the transition from the normal to a non-standard mode will be up to 16 h;
- The sinter plant, rolling production furnaces, and repair shops either stop operating or switch over to a mixture of combustible gases composed of blast furnace gas and coke oven gas only, within about 2 h; within the same 2 h, the CHP plant is switched over to a backup fuel and, the level of natural gas supply to the blast-furnace shop is reduced by about one-half; as a result, within the first 2 h of an

emergency in the natural gas supply system, its supply to the steel mill can be reduced by about 40%;
- the natural gas in the blast furnace process gets substituted completely for the corresponding amount of coke in about 4 h (it should be reminded that the blast furnace process in the integrated steel mill uses up to 5% of the natural gas supplied to the steel mill under its normal operation mode);
- the mode of natural gas supply to the open-hearth shop (up to 35% of the volume of gas used at the plant under normal conditions) does not change in the event of an emergency.

Arriving, in a substantiated way, at a reasonable hourly schedule of natural gas supply to the steel mill in the event of its transition to a non-standard mode of operation requires a review of such operation and process specifics of gas consumers of the mill (in addition to those features that were reviewed above), such as dependence of output volumes of their products by ironworks on the level of undersupplies of natural gas to them and the maximum depth of a reduction in gas supplies, at which the steel mill stops production of its products.

Dependence of ironworks on the level of undersupply of natural gas to them. The sinter plant proves to be most cost-efficient when using up to 30% of natural gas in a combustible mixture. When natural gas is substituted for a mixture of coke gas and blast furnace gas, the heat treatment of iron ore raw materials is disturbed and then the output of pellets may decrease by 20–30%.

Coke oven batteries are very demanding in terms of the quality of gas supplied to the burners of heating chambers, especially with regard to its heat of combustion. Here, the substitution of natural gas in the mixture of gases used is not recommended during the planned "baking" of a coke charge (i.e., when the onset of the emergency coincides with the baking of the currently processed charge), otherwise, the products (coke) will be defective. Of course, all of the above only applies to those coke oven chambers where the coking process is active. Substituting one mixture of combustible gases for another is possible only for a chamber where the charging has started for subsequent coking of the charge. And since such charging proceeds from one chamber to another, the substitution of one fuel gas composition for another also proceeds in the same order. Of course, the combustion heat of the new fuel mixture (free of natural gas) will be lower (than that of the mixture that contains natural gas). This will lead to some increase in the baking time of the coke oven charge in each chamber, which, in general, will reduce the performance of the coke oven battery to about 10% (with the complete rejection of the use of natural gas for heating of coke oven chambers).

Blast furnace process. To intensify the process and reduce coke consumption, natural gas is fed through the blast furnace tuyeres. The blast furnace gas is enriched with natural gas to increase the dome temperature of stoves and their overall efficiency. The gas mixture consumption by stoves remains relatively constant during their operation. The blast furnace process depends on the efficient operation of stoves, therefore, blast furnace stoves are among the main protected gas consumers. When natural gas is not used during an emergency and the gas is substituted with a certain amount of coke, the intensity of iron melting decreases slightly. As a result, the cast iron melting time will increase slightly (or the blast furnace performance will decrease by 5–7%).

Steelmaking facilities (open-hearth furnaces and basic-oxygen converter production) are an integral part of the integrated steel mill production, the termination of gas supplies to them (or a noticeable reduction therein) may lead to a complete shutdown of the integrated steel mill with huge negative consequences for the country's economy (including shutdown of blast furnaces and coke oven batteries). As mentioned above, this option is simply not acceptable: natural gas should be supplied to the open-hearth shop during emergencies in the same full volume as when operating under normal conditions (about 35% of the full volume of natural gas consumption under these conditions). From the above it follows that the *maximum depth of a reduction in natural gas supplies during emergencies to the fully integrated steel mill*, at which *the plant completely ceases to operate* is about *30–35% of the supply under normal conditions of gas supply*, i.e., in this case, the operation of steelmaking is deemed essential.

Rolling and heat treatment of metals. The metal is heated in furnaces of various designs prior to entering rolling mills and during heat treatment. These furnaces are heated with natural gas, coke oven gas, blast-furnace gas, or mixtures thereof. On average, it can be assumed that out of 100% of gas mixture consumed by the rolling production blast furnace gas accounts for 40%, 40% is coke gas, and 20% is natural gas. Some consumers are able to reduce their consumption of natural gas by changing the process of rolled metal production. Others can switch over to heating by a backup fuel. The output of rolled products given the substitution of natural gas decreases by (10–20%).

Auxiliary and buffering processes. Auxiliary consumers (linear, forging, thermal, etc.), CHP plants, and heating boiler houses have the possibility to switch over from one type of fuel to another and the volume of their production output does not depend on the level of gas undersupply.

2.1.3 Algorithms of the Rational Transition of the Integrated Steel Mill from Operating in the Normal Mode to Operating in a Non-standard Mode and Back to the Normal Mode

In this case, *the algorithm of the rational transition from the normal to a non-standard mode* is essentially *minimum required hourly volumes of natural gas supplies* to the specified mill during the time period of its transition from the normal to a non-standard mode without causing serious economic damage to this mill and the country as a whole. The above time period starts from the moment when the emergency occurs ($t = 0$) and ends with the moment when the operation mode of the steel mill reaches a stable (if non-standard) mode.

The algorithm of the rational transition of the steel mill from a non-standard mode of operation to the normal mode is essentially the required hourly volumes of natural gas supplies to the mill *within the time interval* from *the moment the emergency in the gas supply system is terminated* until the mill restores completely its normal mode of operation. These volumes are determined, on the one hand, by the required volumes of gas supply to the steel mill under normal gas supply and, on the other hand, by the upper limits, which are imposed by the operation and process specifics of main gas consumers of the steel mill during the transition from a backup fuel to natural gas. The algorithm is aimed at the fastest restoration of the normal mode of operation of the steel mill to minimize the economic damage associated with the need for the plant to operate in a non-standard mode during an emergency.

The overall pattern of the rational schedule (the algorithm content) of the transition of the steel mill from its normal mode of operation to a non-standard mode *and the duration* of such a transition, strictly speaking, is determined by *the correspondence* between the moment of the onset of the emergency and those stages of the process that are performed at this moment in coke production and steelmaking. We assume that the blast furnace process together with sinter plants, the rolling mill, the buffer storage CHP plant, and auxiliary production facilities have a uniform hourly schedule of their natural gas supply (under normal operation conditions). The transfer of these consumers to new combustible gas mixtures, fuel oil (for CHP plants), an additional amount of coke (for blast furnaces) can be started immediately (after the onset of an emergency).

With regard to the special aspects of steelmaking, when the open-hearth shop is part of the integrated steel mill, it has already been mentioned above that there is no backup fuel and one cannot shut down this shop during an emergency. Natural gas should also be supplied here during emergencies in the amount equal to approximately 30–35% of the total volume of this gas consumed by the entire steel mill under normal conditions.

At coke oven batteries the situation is different: here one can change the composition of combustible gas mixtures and do away with natural gas. However, this can be done only by proceeding in the chamber-by-chamber fashion, starting with the chamber where the end of the coking process of the next coke charge coincided with the onset of the emergency. A complete cessation of natural gas supply to coke oven batteries requires approximately 15–16 h in this case. *This is the duration of the transition of the entire integrated steel mill from the normal to a non-standard mode of operation.*

The rational schedule (algorithm) of the transition of the steel mill from the normal to a non-standard mode is shown in Fig. 2.4.

Figure 2.4 shows the following:

- stages of the transition of the integrated steel mill from the normal to a non-standard operation mode due to an emergency in the gas supply system (I, II, III, IV), where:

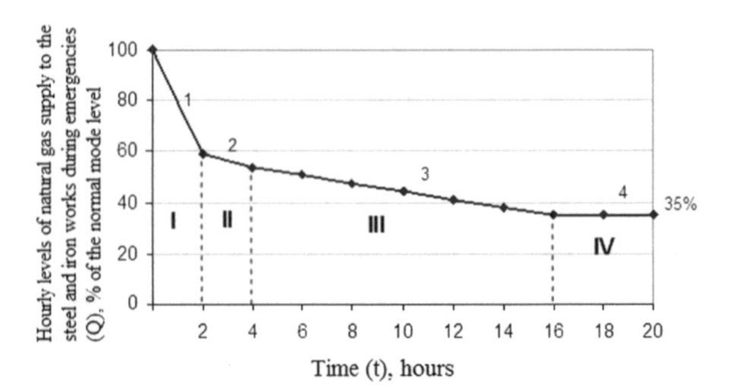

Fig. 2.4 The rational schedule (algorithm) of the transition of the integrated steel mill from the normal mode to a non-standard one due to an emergency in the natural gas supply system ($t = 0$—the onset of an emergency)

I. a reduction (with complete termination) of natural gas supply to the buffer storage CHP plant, sinter plant, rolling mill furnaces, auxiliary production facilities; a reduction in natural gas supply to blast furnace smelting facilities and coke oven batteries;
II. a reduction (with complete termination) in natural gas supply to the blast furnace shop with substitution of this gas for an additional amount of coke;
III. a reduction (with complete termination) in natural gas supply to coke oven batteries; the composition of combustible mixture for heating its chambers changes: the share of coke gas in this mixture increases;
IV. the period of the steel mill operation in a non-standard mode;

- admissible rates of natural gas supply decrease during the transition of the steel mill from the standard mode of operation to a non-standard mode (1, 2, 3, 4), where:

1. overall reduction rates of the above levels at the rolling mill furnaces, sinter plant, buffer storage CHP plant, auxiliary facilities, blast furnace smelting facilities, and coke production facilities;
2. Completion of natural gas substitution for an additional amount coke in the blast furnace process;
3. rates of continuous reduction in natural gas supply to coke oven batteries at the virtually constant level of its supply to steelmaking (open-hearth shop, basic-oxygen/oxygen converter process);
4. the level of natural gas supply to the steel mill after it transitions to the non-standard mode of operation.

The pattern of the rational schedule (the content of the algorithm) of the transition of the steel mill from a non-standard mode of its operation to the normal mode after the emergency is over (or the rational schedule of restoring the regular mode) will be determined by the same operation and process specifics that determine the overall pattern governing the rational schedule of transition of the plant from the regular operating mode to the non-standard one. To build a rational schedule for restoring the normal mode, one can make use of the results of the analysis of possibilities and duration of restoring such a mode for the main gas-consuming facilities of the steel mill. These possibilities and duration depend mainly on one circumstance, that is whether these facilities are located outside of the integrated steel mill or whether the facilities in question are part of such a mill.

Let us first consider the first case, when open-hearth furnaces, coke oven batteries, sinter plants, and rolling facilities are standalone entities outside of the integrated steel mill (blast furnaces that operate outside of such an integrated steel mill are omitted from consideration).

Open-hearth furnaces. The charge gets loaded (most often it is scrap metal). At the very beginning of melting the hourly gas supply volume should be higher than 100% (up to 120%), where 100% is the operation of such a furnace at the beginning of melting under the normal gas supply mode (with no emergency). Here (unlike with the normal mode), it is required to heat not only the charge but also the furnace itself.

Coke oven batteries. Here, as is the case during the transition from the normal to a non-standard mode, the substitution of one combustible mixture for heating of coke oven battery chambers (without natural gas) for some other combustible mixture (with natural gas) should be carried out in a chamber-by-chamber fashion as soon as the "baking" of coke charge is finished in the chambers fired by a combustible mixture with no natural gas. Such substitution begins immediately after the emergency is terminated.

Sinter plants and rolling furnaces can be switched over to a combustible mixture that uses natural gas, along with coke oven batteries, immediately after the emergency is over.

Next, we consider the other case, that is when all gas consumers are part of the integrated steel mill. We assume that at the time when the emergency is terminated, the steel mill was operating with a minimum volume of natural gas supplies to it (as low as 35% of the volume that it consumes under normal gas supply conditions). This 35% allows the open-hearth process to operate normally with the entire mill operating satisfactorily (albeit with some decrease in performance).

From the moment of termination of the emergency in the system of natural gas supply to the steel mill, when only open-hearth process is operating normally if there are opportunities for the complete supply of this gas available, the following is to be carried out:

- substitution of a part of the coke used with natural gas in blast furnaces; due to such substitution alone the natural gas supply to the mill should increase within 4 h by approximately 5% in terms of what is supplied to the plant under normal conditions;
- the transfer within about 2 h to the utilization of combustible gas mixtures with the inclusion of natural gas for the sinter plant, rolling mill furnaces, buffer storage CHP plant, and auxiliary production facilities; by the end of such a transfer the volume of natural gas supply to the mill should increase (due to the transfer of these facilities alone), by about 35–40% of the total supply to the mill under normal conditions;

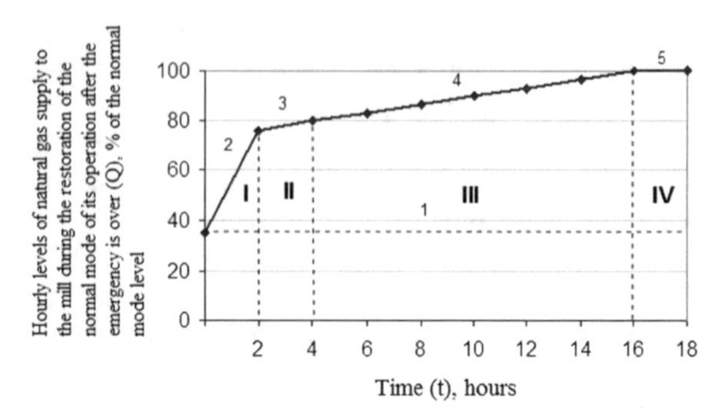

Fig. 2.5 The rational schedule (algorithm) for restoring the regular mode of operation of the integrated steel mill after the emergency in the natural gas supply system is terminated

- the transition of coke oven batteries from a mixture of blast-furnace gas and coke oven gas only to a mixture of these gases and natural gas; the transition is carried out in the order indicated above, the time this transition takes is about 16 h.

Taking into account the above, an outline of the rational schedule (algorithm) of restoring the normal mode of operation of the integrated steel mill will be as shown in Fig. 2.5. The figure indicates the following:

- stages of the transition of the integrated steel mill from a non-standard operating mode (due to an emergency in the gas supply system) to the normal operating mode after the emergency (I, II, III, IV), where:

I. incremental increase of hourly natural gas supply from zero to the normal (regular) level at the rolling mill furnaces, sinter plant, auxiliary production, and buffer storage CHP plant; partial substitution of the "additional" coke[1] for natural gas in the blast furnace process; beginning of the chamber-by-chamber change of the composition of the combustible gas mixture (towards the increase of the natural gas share) at coke oven batteries;
II. the stage of completing the "additional" coke substitution for natural gas in the blast furnace process;
III. the stage of continuing the chamber-by-chamber change of the composition of the combustion gas mixture in the direction of increasing the share of natural gas with bringing this share to the level utilized when operating in the normal mode; the moment of bringing the share of natural gas in the combustible gas mixture supplied for heating of the chamber of coke oven batteries to the level of natural gas corresponding to operating under normal mode: the end of the transition of the entire steel mill to the normal mode of operation;
IV. the operation of the steel mill in the regular mode after the emergency in the gas supply system is terminated;

[1] Additional coke: the amount of coke that was added into blast furnaces to its default part when natural gas supply to these furnaces was stopped.

- rates of increasing incrementally the levels of natural gas supply to the integrated steel mill during the restoration of its normal operation mode after the emergency in the gas supply system is terminated (1, 2, 3, 4, 5), where

  1. the level of natural gas supply to the mill at the moment when the restoration of the normal mode begins (given that during the operation in a non-standard mode, natural gas was supplied only to steelmaking);
  2. the overall hourly schedule of the natural gas supply incremental increase to all gas consumers of the steel mill during its transition back to the normal mode with bringing this supply to the normal (regular) level at the rolling mill furnaces, sinter plant, buffer storage CHP plant, and auxiliary production facilities;
  3. the overall hourly schedule of the natural gas supply incremental increase to the blast furnace process (with completion at this point of the substitution of the additional coke for natural gas), and to coke oven batteries (at constant levels of this supply of natural gas to the steelmaking process, rolling mill furnaces, sinter plant, buffer storage CHP plant, and auxiliary production facilities);
  4. increasing the share of natural gas in the mixture of combustible gases supplied to coke oven batteries, bringing this share to the normal (regular) level (given the constant regular level of gas supply to all other gas consumers of the steel mill);
  5. the level of natural gas supply to the steel mill after restoring its normal operating conditions.

2.2 The Non-ferrous Smelting Industry

2.2.1 The Companies Under Consideration Are Gas Consumers of the Non-ferrous Smelting Industry

The makeup of the total consumption of various types of fuel by non-ferrous smelters is approximately as follows: 45%—natural gas, 35%—coal, 10%—fuel oil, 10%—coke and coke breeze.

The largest consumers of natural gas in the non-ferrous smelting industry are aluminum, copper, lead and zinc, titanium and magnesium industries; in addition, natural gas is widely used in non-ferrous metal processing and secondary non-ferrous metal production. A sudden interruption of energy supply to the above-mentioned production facilities can lead to severe disruption of technological processes, huge losses, and in some cases to fires and explosions. The transfer of processes at non-ferrous smelters from the normal mode to the minimal consumption mode followed by a fail-safe shutdown takes time, the amount of which is determined by the features of the process of the given smelter.

Table 2.1 Relative volumes of natural gas used by non-ferrous smelters

Smelters	Gas volumes (%)
Aluminum	35
Copper	15
Lead and zinc	15
Nickel	10
Titan and magnesium	10
Heat treatment of non-ferrous metals	10
Non-ferrous metal scrap recycling	5
Total	100

Table 2.1 shows the approximate relative volumes of natural gas used at various non-ferrous smelters (in relation to the total volume of its consumption at these plants).

The enterprises for the heat treatment of non-ferrous metals and scrap processing of non-ferrous metals listed in Table 2.1 contribute, of course, to a certain extent to the country's economy, but this contribution is not so great in comparison with the contribution of enterprises producing "the primary metal" (aluminum, copper, lead, zinc, nickel, titanium, and magnesium). Moreover, the first two types of companies mentioned above are of only moderate importance as consumers of natural gas (15% of all gas consumed by the non-ferrous smelting industry). Therefore, companies of the following industries have been chosen for our analysis of the issues of controlling the modes of operation of gas consumers of the non-ferrous smelting industry under conditions of reducing, terminating, and restoring gas supplies:

- aluminum industry,
- copper smelting,
- lead, zinc, and nickel smelting;
- titan and magnesium smelting.

The study of the indicated issues for each type of the enterprise out of those listed above, in general, is carried out in accordance with the procedure outlined in Sect. 1.3 (identification of operation and process specifics of a given type of enterprises, assessment of the importance of their uninterrupted gas supply during an emergency, the development of rational schedules of their transition from the normal mode to a non-standard one and rational schedules or algorithms to restore the normal mode).

As for the degree of importance of uninterrupted gas supply, all the above mentioned non-ferrous smelters can be considered equally important except, perhaps, for the aluminum industry. In case of the latter, a complete shutdown of the enterprise for a relatively long time (the duration of an emergency plus the time it takes to restore damaged fixed assets—in case of a sudden interruption of gas supply) will lead to huge underproduction of products that occupy a prominent place in Russian exports.

2.2.2 Aluminum Industry Enterprises

The main process consumers of fuel in aluminum production are *tubular rotary kilns for sintering and calcination at alumina shops*. The transfer of the kilns to natural gas has allowed to greatly facilitate the stabilization of a thermal mode of sintering of aluminum-containing ores in a range of temperatures from 1150 to 1350 °C and to make its full automation possible.

The use of natural gas is very efficient for alumina plants that process nepheline ore and produce soda ash and potash along with alumina. Sintering of nepheline ore with limestone requires stricter compliance with the furnace temperature profile than sintering of bauxite charges when the temperature fluctuation, at which the sintering occurs, should not exceed 40–50 °C (1275–1325 °C). Therefore, it is almost impossible to maintain the required thermal conditions when running on carbon dust or fuel oil.

The experience of aluminum smelters that use natural gas as a process fuel has shown that it is a reasonably good intensifier of processes. For example, at the Bogoslovsk Aluminum Smelter, after switching over to natural gas, the specific consumption of the reference fuel for calcination of alumina decreased from 178 to 173 kg/t, and for sintering of the charge (when switching over from coal to gas), it decreased from 125 to 120 kg/t. The use of natural gas allows for reducing the cost of alumina production by 5–6%.

Production of fluoride salts. In the production of fluoride salts used for aluminum electrolytic reduction, gas consumers are rotary tube-type furnaces (reaction and drying furnaces). The use of natural gas instead of producer gas in the production of fluoride salts allows increasing the performance of reaction and drying furnaces by 5–6%.

Production of electrode products. Such production covers anode paste, coal, and graphitized products. The main fuel consumers of this sub-industry are coal kilns for pressed carbon slabs and pre-heat furnaces for all types of products. Only the gas fuel, i.e., natural gas, is used for roasting furnaces. As an exception, fuel oil is used in calcining furnaces only at the plants located in areas where there is no natural gas available.

At those enterprises of the aluminum industry that have a system of supplying them with natural gas, *such gas is supplied to the plant's CHP plant as well*. The CHP plant in this case is needed to produce steam of various parameters. This steam is required for alumina production.

Aluminum is produced by electrolysis of alumina dissolved in electrolyte consisting of molten cryolite with addition of some salts. Three main types of electrolyzers are currently in use in the aluminum industry: with the self-baking coal-paste anode and horizontal steel studs carrying an electrical current of 60–90 kA, with the self-baking coal-paste anode and vertical steel-aluminum studs carrying an electrical current of 100–180 kA, and with baked anodes made from several carbon blocks and horizontal studs, embedded in the blocks, carrying an electrical current of 45–250 kA.

The process of aluminum production is continuous and among other non-ferrous smelters, it consumes natural gas the most—more than one-third of the total volume of gas that is supplied to all non-ferrous smelters (Table 2.1). If the supply of natural gas to the (aluminum) smelter is interrupted without an emergency shutdown, the sintering and calcination furnaces of alumina workshops and the furnace for production of fluoride salts can be switched over to other fuel for a certain period of time. *Only the production of electrodes does not allow for substitution of natural gas for other fuel during an emergency.*

A decrease in natural gas supply during emergencies at the aluminum smelter will naturally affect the output of this enterprise. First, the substitution of natural gas for coal dust or fuel oil by tubular rotary kilns of sintering and calcination of alumina makes it difficult to stabilize the thermal profile of the sintering of alumina-containing ores, which reduces the performance of such kilns by 7–10%. Second, when natural gas is substituted by other fuel at the furnaces for the production of fluoride salts, there is also a certain decrease in the output of products (the salts) by 5–7%. To sum up, it can be said that by setting the natural gas supply to the aluminum smelter during an emergency at about 30% of the level of this plant's normal operation, the plant will continue to operate, but with a smaller output (a decrease if compared to the normal mode may be up to 10%). It should be reminded that the aluminum smelter in this case, in addition to the pots for direct metal smelting, has the alumina shop, the fluoride salts production shop, and the electrode production shop. The above 30% of gas supplied to the smelter during an emergency is used only in the production of electrodes (where it is impossible to substitute gas for other fuel). And the same 30% should be deemed as the limit of gas supply levels when further a reduction in the level leads to a complete shutdown of the plant.

When developing a rational schedule (algorithm) of the transition of the given enterprise of the aluminum industry from the normal mode of operation to a non-standard one (because of the emergency in the gas supply system) one should take into account the following operation and process specifics of this smelter:

- substitution of natural gas for some other fuel at the sintering and calcination furnaces of alumina production is allowed; by such substitution alone, in a matter of approximately 2 h after the onset of the emergency, one can reduce the gas supply to the smelter by 40–45%;
- within 3 h (starting from the onset of the emergency) it is allowed to substitute natural gas for other fuel at the furnaces producing fluoride salts: this substitution reduces the level of gas supply by another 25–30%;
- natural gas is supplied to the electrode production shop during emergencies in the same volume as under the normal mode.

The specified procedure (algorithm) of the transition of the aluminum smelter from the normal mode of operation to a non-standard mode enables one to continue the operation of the smelter with a slightly reduced performance (a decrease can be up to 10%) and with relatively small economic losses.

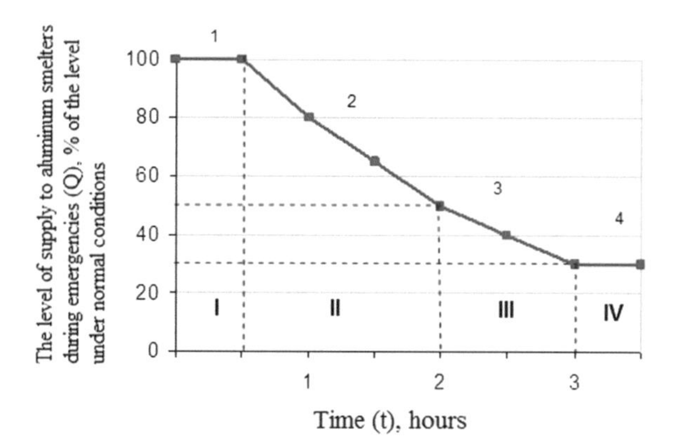

Fig. 2.6 A rational schedule (algorithm) of the transition of the aluminum smelter from normal operation to a non-standard mode during an emergency in the gas supply system ($t = 0$—the onset of an emergency)

Figure 2.6 shows a rational schedule (algorithm) of the transition of the aluminum production company from the normal to a non-standard mode (without stopping the operation of the smelter). The figure indicates the following:

- the stages of the transition of the smelter from the normal to a non-standard mode according to the levels of natural gas supply it demands (I, II, III, IV), where
  I. the stage of preparation of furnaces of sintering and calcination of alumina as well as furnaces for the production of fluoride salts so as to make them run on a new fuel;
  II. the stage of a reduction in natural gas supply levels in sintering and alumina calcination furnaces with complete cessation of its supply to these furnaces by the end of the second hour, as well as a partial reduction in gas supply levels for the production of fluoride salts;
  III. a continued reduction in natural gas supply to fluoride salts production with complete cessation of the supply by the end of the third hour starting from the onset of an emergency;
  IV. the operation of the smelter in a non-standard mode;
- the minimum allowable hourly levels of natural gas supply to the aluminum smelter at different (as indicated above) stages of the transition of the smelter from the normal to a non-standard mode (1, 2, 3);
- the minimum allowable level of gas supply to the given smelter during its operation in a non-standard mode without stopping the smelter.

Figure 2.7 presents a rational schedule (algorithm) of restoring the normal operation mode of the aluminum smelter for the case when this smelter did not stop its operation during the emergency, i.e., it kept operating with the supply of natural gas to the smelter in a non-standard mode in the amount of about 30% of the normal level. The figure indicates the following:

- stages of the transition of the aluminum smelter from a non-standard mode back to the normal mode (restoration of the regular mode)—I, II, III, IV, where

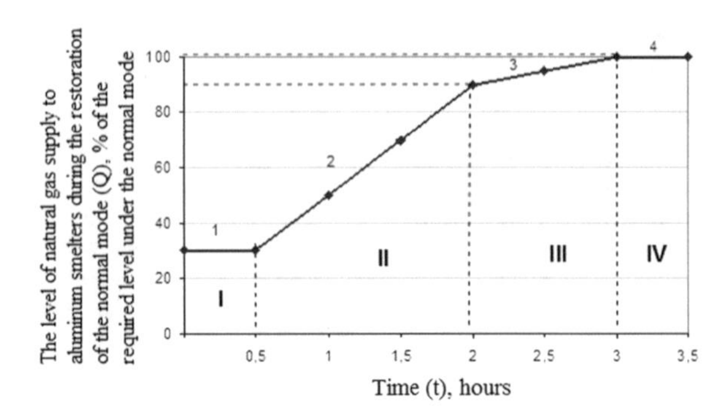

Fig. 2.7 A rational schedule (algorithm) of restoring the normal operation mode of the aluminum smelter after the emergency in the gas supply system is terminated

I. the stage of preparation for the transition of the smelter's facilities to natural gas;
II. an increase in gas supply levels for sintering and calcination furnaces of alumina with the maximum output of these furnaces by the end of the second hour; an incremental increase of gas supply levels for fluoride salts production;
III. a continued increase in gas supply levels for the production of fluoride salts with completely restoring normal operation of these furnaces and the entire smelter;
IV. operation of the smelter in the normal mode (after the emergency is terminated);

- hourly levels of natural gas supply during the above stages of the transition from a non-standard mode to the normal mode.

2.2.3 *Copper Smelters*

The copper smelting process is continuous. Natural gas is used there to dry ores and concentrates, as well as in reverberatory melting furnaces and for copper refining. Out of 100% of natural gas consumed by the copper ore refining facility, about half is used to dry ores and concentrates. The other half of the gas is burned in reverberatory furnaces and during copper refining. Drying of copper ores and concentrates is carried out in rotary tube dryers or in fluidized bed plants. The use of natural gas makes the drying process more controlled and adjustable, which allows increasing the units' output by 30–40% and reducing specific fuel consumption by 2–3%. The roasting of copper ores and concentrates is also carried out in fluidized bed furnaces. In the course of reverberatory melting, the use of natural gas is process-wise combined with the use of oxygen and hot blast, which allows to significantly intensify the melting process.

Drying of ores and concentrates provides for the possibility (in case of a shortage of natural gas) to switch over dryers to other fuels. However, as a result of such substitution generally the drying process gets drastically worse and with the decrease

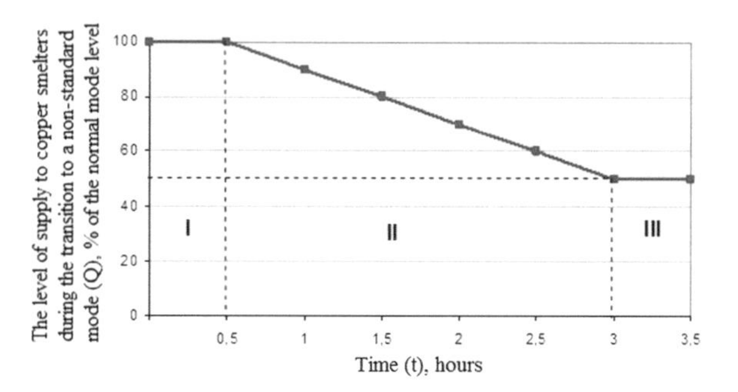

Fig. 2.8 A rational schedule (algorithm) of the transition of the copper smelter from the standard mode to a non-standard one due to an emergency in the gas supply system, without stopping the operation of the plant ($t = 0$—the onset of an emergency)

in the performance of dryers by 30–40%. Correspondingly, the output of finished (end) products by the copper producer also drops.

Reverberatory smelting and refining processes do not allow the substitution of natural gas for any other fuel (and these processes take up about 50% of all gas consumed by the smelter under normal operation). Therefore, a reduction in natural gas supplies to the plant to the level below 50% results in a shutdown of the plant, and since the copper smelting process is continuous, a sharp drop in gas supplies may also lead to the destruction of furnaces of copper reverberatory smelting and refining.

Figure 2.8 shows a rational schedule (algorithm) of the transition of the copper smelter from the standard mode of operation to a non-standard one due to an emergency in the gas supply system.

Figure 2.8 shows the following:

- stages of the transition of the copper smelter from the normal mode of operation to a non-standard one during an emergency in the gas supply system (I, II, III), where

  I. the stage of preparation of devices used for drying of copper ores and concentrates aimed at making them ready for the transition to a new fuel (to substitute for natural gas);
  II. gradual substitution of natural gas with a backup fuel at the units for drying of copper ores and concentrates with complete substitution by the end of the third hour from the onset of the emergency; the smelter will enter a non-standard mode of operation by the end of the same (third) hour;
  III. operation of the smelter in a non-standard mode during an emergency;

- hourly levels of natural gas supply to the copper smelter at different (those indicated above) stages of its transition to a non-standard operation mode.

A rational schedule (algorithm) of restoring the normal mode of operation of the copper smelter after the emergency is over will actually mirror the schedule in Fig. 2.8, i.e.,

- at the first stage, about half an hour is spent preparing the units used for drying copper ores and concentrates to run on natural gas; during this time, the smelter (if

it was not shut down completely) is supplied with 50% of the gas volume required under the normal operation mode (here, the gas is used for copper melting in reverberatory furnaces and its refining);
- then (stage two) during the next 2.5 h a backup fuel is gradually substituted for natural gas at the ore and concentrate drying units with the recovery of the normal operation mode of these units and the smelter as a whole.

2.2.4 Lead, Zinc, and Nickel Smelters

Being consumers of natural gas, the lead, zinc, and nickel smelters are to some extent identical, although the processes employed by these smelters differ markedly.

Lead smelting. Raw materials for lead smelting (as available at the smelter) are concentrates produced at the site of lead ore mining using the flotation method of beneficiation. Concentrates may contain 50–80% of pure lead. Recovery of lead from such concentrate is carried out by the pyrometallurgical method:

- sinter roasting of lead concentrates;
- reduction smelting of sinter in shaft-type furnaces;
- refinement of pig lead; along with the refinement process, processes for recovering impurities (as useful products) are performed.

Agglomeration sintering is carried out with belt sintering machines with partial removal of sulfur and obtaining the lump product. The main fuel is natural gas.

An important lead smelting process is reduction blast smelting. Fuel for reduction blast smelting is blast furnace coke, the consumption of which is 8–15% of the mass of the melted charge. Recently, natural gas has been used for heating of blast and as a partial substitute for coke. The melting process results in pug lead that has various impurities (copper, arsenic, zinc, bismuth, tin, antimony, gold, silver, etc.) concentrated in it. Pig lead is then refined and purified from impurities. The main technique used to refine lead is that of the pyrometallurgical method that is implemented in small-capacity machines that require a significant amount of labor, energy, and chemicals. Refining units are switched over to heating by natural gas instead of coal and fuel oil.

Slags containing a number of useful impurities and semi-products are formed in the process of reduction melting. Slags are therefore processed (to extract useful impurities) using the fuming technique, which is blowing molten slag with a mixture of natural gas and air (here natural gas is needed as a reducing agent).

Lead smelting allows for its complete shutdown in the event of a gas supply interruption. However, it takes about 3 h since the onset of the emergency in the gas supply system for a complete fail-safe shutdown to take place.

If the auxiliary fuel handling facilities have been preserved, it is possible to switch over some of the gas consumers of the lead smelter to fuel oil or coal without stopping the smelter. However, no-shutdown operation during the emergency requires the supply of natural gas to the plant during this emergency as well (about 50% of the level of supply maintained under normal conditions).

Restoring the normal mode of operation of the lead smelter after the emergency in the gas supply system is terminated, after the complete ceasing of operation, requires about 8–10 h (starting up the plants and units from the cold state), and the complete switchover of the plant (in the case when there was no shutdown) to natural gas takes about 3 h.

Zinc smelting. After beneficiation by flotation, the zinc smelter receives as its raw material a concentrate with zinc content ranging from 20 to 50%.

Zinc from zinc concentrates is obtained either by the hydrometallurgical method or by distillation. The essence of the hydrometallurgical method is as follows. The concentrate that comes from ore-dressing mills is roasted in fluidized bed furnaces at 900–950 °C using the heat of exothermic reactions. The roasted concentrate is leached with a weak solution of sulfuric acid to obtain zinc sulfide in mixers with air or mechanical mixing. After leaching, the solution is cleaned of impurities, neutralized, and directed to the electrolytic recovery of zinc. Existing multistage workflows of zinc concentrates processing at their individual stages of processes yield various intermediate products and semi-products. Some products are returned to the charge for further processing, while some, such as zinc cakes from the hydrometallurgical process, are removed and processed separately. The most common way to process zinc cakes is the Waelz process carried out in rotary tube furnaces at 1000–1200 °C. Coal and coke are introduced into the charge during the Waelz process.

When zinc is obtained by distillation, the roasted concentrate is loaded into a retort oven made of fire clay paste. For heating of the furnace, one mainly uses natural gas. The total extraction of zinc during the distillation process is 90–93%.

Zinc smelters in case of an emergency in the natural gas supply system allow their complete fail-safe shutdown. It takes about 3 h. If secondary fuel systems are available (e.g., fuel oil), it is possible not to stop the operation of this smelter. However, this cannot be done if zinc production is carried out as per the method of distillation, when retort ovens are used for roasting zinc cakes, which do not allow for the substitution of natural gas for other fuels. These retort ovens can consume up to half of all the gas supplied to the entire plant under its normal mode of operation. The supply of this half should be preserved during the operation of the smelter in a non-standard mode as well.

Just like with lead smelters, the cold start of the zinc smelter takes up to 8 h, while taking approximately 3 h when starting from the running state.

Nickel smelting. In the smelting of nickel and associated metals, natural gas is used for drying and roasting of ores and concentrates, for incinerators during agglomeration of concentrates, for roasting of nickel matte in fluidized bed furnaces, for slag depletion in converters, and for heating of the blast in shaft kilns.

The high heat of combustion, the recoverability and purity of natural gas allow its efficient use not only for existing processes in this branch of the sector, but also for the introduction of new ones, such as flash smelting of nickel concentrates, smelting with the use of gas-oxygen burners, ore electric smelting with the use of gas-oxygen burners.

The nickel production process is continuous. Almost all basic elements at the smelter nowadays run on natural gas. During an emergency in the gas supply system, a complete fail-safe shutdown of this smelter will take approximately 3 h. During this time, the natural gas supply to the entire smelter should be maintained at 70% of the normal level.

It takes 8–10 h to launch nickel smelters from the cold state after the emergency is terminated. First, the gas is fed in the concentrate drying unit and simultaneously the heating of the nickel matte furnaces is started; then the converter furnaces of the depleted slag are heated.

If it is possible to maintain gas supply to the nickel smelter at the 70% level, it can operate normally during emergencies, yet with a reduced capacity (approximately 90% of the rated capacity).

2.2.5 Titanium and Magnesium Smelters

100% of the natural gas that goes to the titanium and magnesium smelter is used only in titanium concentrate drying, roasting, and sintering devices. This concentrate is obtained by beneficiation of titanium-containing ores (rutile and ilmenite) by the gravitational method with magnetic separation.

The method of obtaining primary titanium by the reduction of magnesium tetrachloride titanium has become widespread. This is why magnesium and titanium are most often produced by the same smelter.

Modern titanium smelting is made up of the following processes: ore mining and beneficiation; drying, roasting, and sintering of concentrate, electric smelting of the ilmenite concentrate to produce rich titanium slag; production and purification of titanium tetrachloride; production of spongy titanium; melting of spongy titanium into ingots; the processing of magnesium chloride into magnesium and chlorine.

Titanium-containing slags are smelted in 5–20 MVA electric furnaces. The melting products are titanium slag ($\approx$90% of titanium dioxide) and cast iron whereto up to 95% of iron is transferred. Tetrachloride titanium is obtained by chlorinating rutile concentrate, titanium slag, or synthetic rutile. Ilmenite concentrates are not chlorinated.

In order to obtain pure titanium, magnesium is used: that is to say, the thermal method based on reducing tetrachloride titanium with magnesium. The metallic titanium formed during the reduction process is collected at the bottom of the reactor as a spongy mass impregnated with magnesium and magnesium chloride. The extraction rate of titanium from slag to the sponge is 70–75%. Ingots from spongy titanium are melted in electric arc furnaces with the water-cooled crystallizer. In the production of titanium, the main energy costs are electric power costs.

The main method for obtaining metallic magnesium in Russia is the electrolytic method, that is when beneficiated ores (carnallite, bischophite, or the chloromagnesium concentrate) are subjected to electrolysis.

Carnallite normally contains water and is dehydrated in two stages. The first stage is performed in rotary tube furnaces at 200–230 °C. The second stage, that of carnallite melting, is performed in electric resistance furnaces with subsequent settling in cooling mixers.

When using bischophite to power the pots, it is dehydrated as per a two-stage scheme similar to that adopted for carnallite dehydration.

The magnesium-chlorate scheme is adopted when using raw materials containing magnesium oxide. Here, magnesium oxide is chlorinated in the electric shaft kilns.

Judging by the above, the main energy carrier at titanium and magnesium smelters is electric power. Natural gas, as already mentioned, is used here only for drying, roasting, and sintering of the titanium-containing concentrate. Today, at the smelters currently operating in Russia natural gas is the only fuel used for these processes.

If before the onset of the emergency there was a lot of concentrate readily available at the smelter (after drying, roasting, and sintering), the smelter can operate as usual even after the onset of the emergency (until the concentrate is used up). In the case where there is no stock of the concentrate, the smelter keeps operating as long as all current processes at the various titanium and magnesium smelting plants that use electricity are in progress. After obtaining the final portions of finished goods during the specified (ongoing) processes the operation of the smelter stops until the emergency is terminated.

Restoration of the normal operating mode of the titanium and magnesium smelter after the emergency is terminated, if performed from the cold state, begins with the supply of natural gas to raw materials drying, roasting, and sintering furnaces. It takes up to 4 h for these devices to achieve the normal mode of operation. The same period should be used to prepare all plants powered by electricity to receive the first portions of finished sinter that are needed in the processes of obtaining the finished product.

2.3 Production of Construction Materials (as Illustrated by the Cement Plant)

The main gas-intensive industry of construction materials is the cement manufacturing process. Today this process consumes about 50% of all natural gas that is used in the construction materials industry.

The processes of heat and chemical processing of raw materials are used in cement manufacturing. The main raw materials are the following: limestone, marl, chalk, clay, slags from the metallurgical and power industries. Thermal and chemical treatment of raw materials is carried out in rotary kilns by the wet, dry, or combined methods. Natural gas, fuel oil, coal (carbon dust) can be used as a fuel. At present, the distribution of fuel types in cement manufacturing in Russia is as follows: natural gas—92.4%, fuel oil—1.5%, coal—6.1%. Thus, natural gas is practically the only fuel in the case of Russian cement plants and there is no backup fuel envisioned for the new plants being created.

First of all, natural gas is used for processing of raw materials. Limestone fines and clay are mixed in the ratio of 4:1 and sent to the drying section of the rotary kiln. Due to high temperatures limestone decomposes into lime and carbon dioxide. Within the temperature range of 1,400–1,450 °C lime reacts with other chemical components and gets sintered. The end product of these reactions is cement clinker that is cooled in the refrigerator.

The final stage of cement manufacturing is the grinding of cement clinker in the cement mill to a powder. Mineral components (slag, fly ash, gypsum, etc.) are added when grinding cement clinker to produce various types of cement.

As a rule, when several operating production lines are available at any given time each line (both during drying of the above mixture and burning of this mixture) has its own stage of the process being carried out. It is possible to interrupt the natural gas supply without any serious consequences for the main production assets of the cement plant at all stages of dryer units operation (loading of the above mixture, drying of the mixture, and unloading of the dried mixture) and at two stages of kiln operation for clinker burning (loading of the same but dried mixture and unloading of the finished clinker from kilns). The main stage, that is the clinker burning stage, does not allow natural gas supply to stop abruptly at the kiln where the burning takes place. It follows from this that a fail-safe shutdown of the cement plant due to an emergency in the gas supply system is only possible given a step-by-step shutdown of production lines. The shutdown starts with the line where the "late" stage of clinker burning is active at the moment of the onset of an emergency, i.e., with the line where there is less time left until the end of burning than on other lines, and the shutdown completes with the line where the moment of the onset of an emergency coincided with the earliest burning stage.

In purely practical terms, it can be assumed that the maximum possible fail-safe shutdown time of the entire cement plant or part of its production lines will be determined by the full duration of the clinker burning as required by the process.

The dependence of the production volume on the level of gas supply to the cement plant during an emergency is straightforward. This volume is proportional to the number of production lines for which the specified level is sufficient for normal operation (for example, if there is enough gas to operate half of the total number of lines in the plant, the plant operates at half its production capacity).

Restoring the regular operation mode of the cement plant after an emergency is also carried out through the phased inclusion of production lines into operation.

Damage caused by the undersupply or complete cessation of natural gas supply to the cement plant is determined by the number of production lines not operating during an emergency and is related in this case only to the undersupply of finished products by the given enterprise.

2.4 The Enterprise that Uses Natural Gas as a Feedstock (the Case Study of Methanol and Ammonia Production)

Nowadays, In Russia, the bulk of methanol and ammonia is produced using natural gas (methane) as a feedstock. Such (co-)producing of these products has a common basis process-wise, that is the application of methane conversion in the presence of oxygen.

Of course, there are various processes available for the production of these products separately (as opposed to having a single technological basis), using various raw materials (oil gases, coke gas, and even products of comprehensive coal processing). However, for the present purpose we are interested only in natural gas and its use as *a raw material* and not as a fuel (as opposed to those industrial consumers whose behavior under conditions of reduced, stopped, and restored gas supplies was covered above).

In this section, it was deemed appropriate to consider the behavior of such a gas consumer during an emergency as co-producing of the above products in the process of methane conversion in the presence of oxygen. This conversion produces acetylene (C_2H_2) and the syngas (a mixture of water vapor, hydrogen, carbon monoxide, and carbon dioxide), i.e.,

$$6CH_4 + 4O_2 \rightarrow C_2H_2 + 8H_2 + 3CO + CO_2 + 3H_2O$$

Once acetylene (a finished and rather valuable product in itself) is removed from the resulting mixture by the adsorption method, what is left is the syngas, the general composition of which is approximately the one shown in Table 2.2.

Judging by Table 2.2, the volumetric ratio of hydrogen to carbon oxides during methane conversion in residual gas is approximately 2:1. This ratio makes it easy to perform the process of methanol synthesis (CH_3ON):

$$2H_2 + CO \rightarrow CH_3OH$$

$$3H_2 + CO_2 \rightarrow CH_3OH + H_2O$$

Table 2.2 Composition and volume fractions of components of the syngas (in case of methane conversion in oxygen and after removing acetylene from conversion products)

Component	Volume fraction (%)
H_2	58–62
CO	26–29
CO_2	0.8–4
CH_4	3.5–6
O_2	0.1–0.5
N_2	0.5–2

After the synthesis of methanol what remains is partly unused hydrogen (H_2) and carbon monoxide (CO); nitrogen (N_2) also remains unused after the separation of air into oxygen and nitrogen. Carbon monoxide and hydrogen from the synthesis unit are directed to recovery. Recovery is carried out by the cryogenic method or by washing with selective solvents. Carbon monoxide is further involved in the production of acetic acid, formic acid, ethanol, etc. At the same time, the hydrogen that was left when oxygen was obtained from the air, is used to produce ammonia (NH_3):

$$N_2 + 3H_2 \rightarrow 2NH_3$$

Main stages of co-producing of methanol and ammonia (by conversion of methane in the presence of oxygen):

- the process of methane conversion in the presence of oxygen to produce a mixture of acetylene and the syngas;
- the separation of the above mixture of acetylene, a finished product that is transferred to the warehouse;
- treatment of the syngas (washing it with water to remove traces of the solvent that was used to extract acetylene);
- the elimination of traces of acetylene, olefins, and methane residues by transforming (converting) them into carbon monoxide and hydrogen and then mixing the products of such conversion with the return gas and CO_2; this return gas may be obtained from ammonia production, for example;
- chemical reactions to produce methanol (see above) by using hydrogen, carbon monoxide, and carbon dioxide as the main components of the syngas;
- separation of methanol from the mixture of reaction products (with its transportation to the warehouse of finished products after cooling and condensation of its vapors), as well as that of emissions of CO and H_2 gases for their further use;
- the process of synthesis of hydrogen and nitrogen to obtain ammonia (hydrogen is obtained after the separation of exhaust gases, while nitrogen is left after the separation of air to obtain oxygen); transportation of ammonia to the warehouse of finished products.

The very conversion of methane (the first stage of obtaining the above products), the process of obtaining methanol, as well as the synthesis of hydrogen and nitrogen in the production of ammonia are very sensitive chemical processes that do not allow any changes in the operation mode of those reactors where these processes are carried out. Here, the following is to be ensured:

- the constancy of ratios of those components that are supplied to reactors as raw materials;
- the continuity of removal of products of the corresponding chemical reaction from reactors;
- the constancy of temperature and pressure in the reactors when carrying out the relevant chemical processes; of course, the specified profile parameters will prove

different each time depending on the essence of the chemical reaction and on the subtleties of the processes used (e.g., with or without a catalyst, etc.).

Any violation of the operation mode of this or that reactor will lead to a change in the composition of those products that come out of this reactor, which in this case means substandard products. For example, any change in the operation mode of the reactor where methane is being converted can lead to such a change in the composition of the products leaving the reactor, that these products would not be suitable for obtaining methanol and, in turn, ammonia from them.

It follows from the above that the supply of natural gas for co-production methanol and ammonia should be constant and strictly apportioned. In the event of an emergency in the natural gas supply system of this production process, it will be necessary to stop the production immediately. It is true that there should be no particular problems with process units and various equipment in case of natural gas supply interruption because all products in the reactors considered here are in the gaseous form and they are easily evacuated from the reactors. After the evacuation, the products are separated into individual components as usual during cooling and by condensation of their individual components (e.g., methanol).

If there is only one production line for co-producing methanol and ammonia at the enterprise, any undersupply of natural gas to this production during an emergency causes the shutdown of production of these products. If there are more than one such line, then depending on the level of such undersupply, those lines for which the available supply of gas during an emergency is sufficient keep operating with the supplied amount being identical to that under the normal mode of operation (just like with the cement production, the features of which were detailed in Sect. 2.3).

Restoring the normal mode of operation of the methanol and ammonia plant for production lines that are not operating during an emergency (or starting up the line from cold state) is a multi-stage process:

- starting up air separation plants (oxygen and nitrogen production);
- a gradual increase in natural gas (methane) and oxygen supply to the reactor for heating of this reactor to achieve the volumes of supply of this gas mixture and the temperature in the reactor that are necessary for obtaining acetylene and syngas components in the required proportions;
- simultaneously with heating of the methane conversion reactor there is heating up to required temperatures of other elements of the production line (where it is required) and also preparation (including heating) for operation of reactors for production of methanol and ammonia.

The required time for the complete transition of the entire process of co-producing methanol and ammonia from the cold state to running with the delivery of the first portion of these products of the desired quality takes 5–8 h (the time depends on the specific features of the production line, where elements of different designs can be used).

Chapter 3
Control of Operation Modes of Gas-Consuming Facilities in the Electric Power Industry Under Conditions of a Reduction, Termination, and Restoration of Gas Supplies

3.1 Homogeneous Groups of Gas Consumers in the Electric Power Industry

Gas-consuming facilities in the electric power industry are fossil fuel-fired thermal power plants (TPP), that is to say those of them where the pipeline natural gas is the main or the only fuel. It should be noted that TPPs in most industrialized countries is one of the main sources of electricity generation: so, for example, in Russia, they account for about 70% of the installed capacity of power plants; with the rest being nuclear power plants (NPPs), hydroelectric power plants (HPPs), and non-conventional power plants (wind, geothermal, tidal) that make up an insignificant percentage of the total capacity. Taking into account that the electric power industry produces two types of marketable products—that is electricity and heat, it should also be noted that TPPs in the country as a whole are the most important, though not dominant, source of heat: they account for one-third of heat produced in the country; the rest is contributed by boiler houses of district heating and distributed heating sectors, individual heat generators, heat recovery units, and installations that run on renewable energy resources. However, TPPs play a crucial role in the heat supply of most major, large, and medium-sized cities and industrial hubs.

The share of natural gas in the mix of the TPP fuel mix in Russia as a whole is more than 68%, and in the European part of the country it makes up 75–95%, including more than 92% in the Central Federal District (CFD), more than 96% in the Volga Federal District (VFD), more than 85% in the Southern Federal District (SFD) and North Caucasian Federal District (NCFD) together, 78% in the Northwestern Federal District (NWFD), and about 75% in the Ural Federal District (UFD). In contrast, in the Siberian (SFD) and Far Eastern Federal Districts (FEFD) the share of gas in the TPP fuel balance has recently been only slightly over 12% and over 20%, respectively.

For the sake of clarity, by gas-fired thermal power plants (TPP) in Russia we mean all the relevant facilities: both public power plants owned by energy (generating and otherwise) companies and power plants owned by other owners, primarily industrial companies, that is—captive power plants and isolated power plants. From what has

V. I. Rabchuk et al., *Control of Operation Modes of Gas Consumers in the Event of Gas Supply Disruptions*, https://doi.org/10.1007/978-3-030-59731-3_3

been said, it is clear that not only power plants that are part of national energy systems are included in our study, but also the plants that are not connected to these systems (electric generation systems) or that operate in a stand-alone fashion.

Taking into account operation and process specifics of different types of gas-fired TPPs, the degree of their responsibility for reliable energy (electricity and heat) supply to consumers and for prevention of unacceptable damage from disruptions of normal energy supply, this study provides for division of gas-fired TPPs into homogeneous groups as based on the following criteria:

- single-product (electricity) or two-product (electricity and heat) facilities with the products being marketable products (in what follows single-product plants are called CPPs, which stands for "condensing" power plants; while two-product plants are called CHP plants, which stands for combined heat and power plants proper;
- the gas-consuming facility based on the steam-turbine (sometimes also referred to as heat-power), steam-gas, or gas-turbine technology;
- the facility is or is not equipped with a backup (or emergency) fuel system; the availability of such a system implies a possibility of the complete switchover from gas to a backup fuel in the event of gas supply disruptions;
- in the case of single-product facilities: whether a facility is burdened or not burdened with the necessity (task) of supplying heat to the settlement adjacent to the plant (or some other residential area adjacent to the plant), for example, by means of heat supply by non-controlled (regenerative) steam extractions from condensing turbines or from steam boilers through a pressure-reducing and desuperheating station (PRDS). In what follows these varieties are referred to as burdened or not burdened with the secondary task.

One last clarification is due: by a gas-consuming facility, we mean a power plant. However, in the case of power plants with complex composition, such a facility may be a part of a power plant: a single unit (module) or a group of units (for example, a district heating unit at a CPP or a gas turbine power unit at a steam turbine plant).

Thus, in the electric power industry we identify the following 18 homogeneous groups of facilities that are consumers of pipeline natural gas (gas-fired TPPs).

1. Steam turbine condensing power plants with a backup (emergency) fuel oil system, not burdened with the secondary task.
2. The same steam power plants (as in item 1) but burdened with the secondary task.
3. Steam turbine CPPs, where a backup fuel oil system was not built or was dismantled for some reason, not burdened with the secondary task.
4. The same CPPs (as in item 3) but burdened with the secondary task.
5. Gas turbine units (power plants) (GTU, GTP) of the "condensing" type (that is, emitting combustion products into the atmosphere), having diesel (or other distillate) fuel as an emergency (backup) fuel and the corresponding liquid fuel system, not burdened with the secondary task.
6. The same GTU (GTS) (as in item 5) but burdened with the secondary task.

7. The same GTU (GTS) as in Group 5, but without a liquid fuel system.
8. The same GTU (GTS) as in Group 7, burdened with the secondary task.
   Note. It should be noted that as of 2012 and for the foreseeable future (until 2020) gas turbine plants of the "condensing" type (Groups 5–8) are not undergoing any large-scale development, unlike with steam turbine CPPs, yet play an important role in providing capacity to cover severe peak loads of electric power systems and are instrumental in providing the system redundancy of electric power capacity, so ensuring sustainable gas supply to them is of utmost importance.
9. Combined cycle gas turbine units and power plants (CCGT units and CCGT plants) of the condensing type (i.e., without waste heat utilization) of all modifications (hereinafter referred to as CCGT-CPP) with a backup (emergency) liquid fuel system and not burdened with the secondary task.
10. The same CCGT-CPPs as in Group 9 but burdened with the secondary task.
11. The same CCGT-CPPs as in Group 9, but with no backup liquid fuel system and not burdened with the secondary task.
12. The same CCGT-CPPs as in Group 11 but burdened with the secondary task.
    Note. To date, the CCGT-CPPs have not undergone any large-scale development, but their construction is considered the mainstream direction of CPPs development in the European part of Russia, so the analysis of their behavior during emergencies in the gas supply system is also extremely important.
13. Steam-turbine CHP plants: steam-turbine power plants with units of the district heating type, that, in addition to electricity, supply heat in the form of hot water and (or) steam both for heating and hot water supply needs of housing and community amenities and (or) industrial consumers, and (or) for process needs of industrial consumers, equipped with turbines of T, PT, R, PR, and TR types, steam and peak hot water boilers having a backup (emergency) fuel oil system.
14. CHP plants similar to those covered by Group 13, but with no fuel oil backup system (for the reasons detailed in the overview of Group 3).
15. Gas turbine CHP plants (GTU-CHP plants) of the same purpose as steam turbine CHP plants (Group 13), including those revamped as GTU-CHP plants by constructing a gas turbine superstructure, district or industrial gas boilers with a backup (emergency) liquid fuel system.
16. The same GTU-CHP plants as in Group 15 but with no backup liquid fuel system.
17. Combined cycle gas turbine CHP plants (CCGT-CHP plants) of different modifications that serve the same purpose as steam turbine CHP plants (Group 13), including the CHP plants revamped as CCGT-CHP plants, industrial or district boiler houses with a backup (emergency) liquid fuel system.
18. The same CCGT-CHP plants as in Group 17, but with no backup liquid fuel system.

3.2 Special Aspects of the Behavior of Electric Power Industry Facilities Under Conditions of a Reduction, Termination, and Restoration of Gas Supplies

A number of studies deal with the issues of changing the operation modes of facilities of the electric power industry under conditions of gas supply disruptions. So, for example, [26] covers the issues of realization of the electric power network potential under conditions of adoption of new technologies by users, especially those related to changes in the energy consumption mode, including that under conditions of energy resource constraints. That having been said, consumers are less afraid of the anticipated additional costs, the more the network's intelligent features reveal themselves so as to mitigate possible negative consequences from interruptions in energy supply.

Elements of facilities of the electric power industry whose operation is impaired when gas supplies are reduced or terminated. For steam turbine CPPs (Groups 1–4), the elements directly responding to changes in gas supply are steam boilers; for steam power CHP plants (Groups 13–14), such elements are both steam boilers (steam generators) and peak hot water boilers; for gas turbine power plants of the condensing and heating types (Groups 5–8, 15–16) such elements are combustion chambers as well as peak hot water boilers; for CCGT-CPP and CCGT-CHP plants and CCGT-CHP plants (Groups 9–12, 17–18) such elements are combustion chambers and steam generators (both high pressure and low-pressure ones) as well as peak hot water boilers. A reduction in gas supply within adjustment range limits, i.e., usually up to 40–60% of the rated flow rate, entails a corresponding decrease in steam (heat) generation capacity of the indicated elements (unless gas is substituted for other fuel), and, as a consequence, a decrease in electricity and heat-generating capacity of steam or gas turbines in accordance with their consumption (energy) performance. A deeper reduction in fuel (gas) consumption or, even more so, an abrupt complete halt of gas supply is unacceptable as they lead to an emergency. In order to avoid this during the planned shutdown of the electric power unit with the termination of gas supply the following should be done: after the gradual unloading (the unloading rate is regulated individually for each type of the power unit, and sometimes individually for each particular unit) to the specified minimum safe output (40–60% depending on the capacity, parameters and technical features of the power unit subject to the shutdown: its boilers, turbines, combustion chambers, and the unit as a whole) it shall be kept in operation at this minimum output for a certain period of time determined by regulatory documents with the same factors taken into account (from 0.5 to 1 h for the GTU to 1–2 h for a steam turbine or combined cycle unit), and sufficient to prepare all systems of the power unit for shutdown, after which the shutdown is carried out, the procedure of which is regulated by the relevant guidelines.

The switchover of the power plant from gas to other fuel and back: its possibility and duration. For gas-fired *steam-turbine* thermal power plants (CPPs and CHP plants: Groups 1–4, 13–14), fuel oil (burner fuel oil or, in some specific cases, bunker fuel oil) is practically the only feasible backup (emergency) fuel and (in most cases, for both operating plants and those at the design stage), in fact, it is the only fuel. In principle, crude oil can be used instead of fuel oil. Back in the 1960s–1990s, steam turbine power plants were equipped with gas- and fuel oil-fired boilers (steam generators), and fuel transport and boiler shops of these power plants were equipped with a gas and fuel oil system. Even if TPPs were initially intended to run on natural gas only, the building codes and regulations (SNIPs) and technological design codes (NPSs) mandatorily provided for backup (emergency) fuel oil facilities. Nowadays, the situation remains basically the same, however, the specified reference documents have lost the legally binding status, therefore when constructing new steam-turbine TPPs (units, modules) and reconstructing (modernizing) existing ones it is not impossible for the investor (or the owner of the TPP), being guided by actual long-established uninterrupted gas supply, to refuse to construct (or even to preserve) a fuel oil system. Thus, groups of gas consumers equipped with fuel oil facilities (Groups 1, 2, and 13) and groups not equipped with fuel oil facilities (Groups 3, 4, and 14) are formed within the overall stock of gas-fired TPPs. With respect to the former, it is possible, in case of an emergency in the gas supply system, to have an essentially smooth switchover from gas to fuel oil (or crude oil) and, naturally, the reverse switchover, while for the latter such switchover is impossible.

As for the duration of the switchover transition of the power unit (module) from gas to fuel oil, this procedure may technically take from 40 to 60 min, depending on the power unit capacity, its condition, as well as the initial condition of the fuel oil system and other circumstances. Approximately the same is the duration of the reverse switchover from fuel oil to gas after an emergency. But the above figures refer to the procedure of switching over from gas to fuel oil of a single unit of the power plant. In case of necessity of simultaneous switchover of all modules (power units) of a multi-unit TPP, taking into account the limited number of highly qualified operators, appropriate instrumentation, other resources, as well as the need to meet, during the entire switchover period, the dispatching electricity and heat load schedules, the duration of the switchover of such a plant from gas to fuel oil (and back) increases to about 1.5–2.5 h.

For *gas turbine* units, gas turbine plants that run on gas (GTU-CPPs and GTU-CHP plants: Groups 5–8 and 15–16), diesel fuel (or some other distillate motor fuel) is practically the only feasible backup (emergency) fuel and (in most cases), it is the only fuel. Fuel oil, even low-sulfur one, contains impurities of sulfur and molybdenum that are unacceptable for the gas turbine performance. However, the equipping of gas turbine units, which were commissioned in the 1990s and are being commissioned nowadays (until then their commissioning had been extremely limited), with a corresponding backup (emergency) liquid fuel system has not been the case (and remains so). Therefore, there are groups of GTUs (GTSs) equipped with a liquid fuel system (Groups 5, 6, and 15) and those not equipped with it (Groups 7, 8, and 16). For the latter, of course, the possibility of switching over to other fuel (if gas supplies are reduced or stopped) is ruled out: they should either be stopped in a fail-safe way (there are appropriate guidelines on how to achieve this) or the minimum required gas supplies must be retained. When a liquid fuel system is available, technically gas turbine unit switches over from gas to liquid fuel in

25–35 min, approximately the same time it takes for the reverse switchover to happen. However, in cases of switching over from gas to liquid fuel simultaneously for all units of a multi-unit gas turbine power plant (or all units of a TPP, where GTUs are installed along with steam turbine units), the indicated duration is increased by as much as 60–120 min for the reasons stated above for steam-turbine TPPs.

This is roughly the same in the case of *combined cycle gas turbine* plants. Those of them that are not equipped with a backup (emergency) liquid fuel system (Groups 11, 12, and 18), naturally, cannot be switched over, in case of an emergency in the gas supply system, from gas to other fuel and are subject in this situation either to a fail-safe shutdown, or to be kept operating with the minimum safe output with the allocation of appropriate gas resources for them.

The availability of a backup (emergency) fuel system (Groups 9, 10, and 17), on the other hand, allows gas consumers (CCGTs, CHC plants) to switch over to liquid fuel smoothly. However, there will be some differences with respect to these subgroups of CCGTs. Those CCGT designs that provide for burning auxiliary fuel in the furnace chamber, can only be expected to switch over to either backup diesel (or other distillate) fuel, or to the same fuel for gas turbines plus fuel oil for additional combustion in the furnace of a steam generator (steam boiler). For those CCGT designs, that do not provide for burning auxiliary fuel, they should switch over only to distillate fuel when gas supplies are terminated. However, the technically feasible duration of the switchover from gas to liquid fuel and the reverse switchover will in any case be approximately the same, that is, falling within the range of 35–50 min. For multi-unit combined cycle power plants, the actual duration of simultaneous switchover from gas to liquid fuel and the reverse switchover of all units will take 75–120 min (the greater numbers in both cases are stated for large combined-cycle units).

Dependence of gas consumer output (deliveries of electricity and heat by gas-fired TPPs) on the level of gas undersupply. Gas-fired TPPs of steam turbine, gas turbine, steam-gas, combined cycle, condensing, and heating types can be operated in an accident-free way within their adjustment range limits, and its lower boundary, i.e., the minimum safe output, is the value of 40–60% of the rated capacity (depending on the type of power plant, unit capacity, and parameters of the power plant, the special features of its design, and the characteristics of individual elements). The minimum safe output is an individual characteristic of each specific module (power unit), and the corresponding fuel consumption (gas supply) is that minimum consumption at which the performance of the power plant is still maintained, that is its output of electricity (for CHP plants—Groups 1–12) or electricity and heat (for CHP plants—Groups 13–18). This minimum gas consumption, which is approximately 40–60% of the rated (normal) one, is descriptive of the maximum depth in a gas supply reduction, at which the gas-fired TPP stops “delivering” its products.

In case of a decrease in gas supply (undersupply) within adjustment range limits of the power unit (of the condensing or heating type) from 100% to the minimum safe output (40–60%) the power output (supplied electricity) of a condensing unit or the total power output (supplied electricity and heat) of a heating unit decreases by approximately the same amount. In other words, there is a linear or close to the linear dependence of energy output ("production output") on fuel consumption (gas supplies). However, the situation is somewhat idiosyncratic when it comes to steam turbines and, to some extent, to combined cycle CHP plants. By means of adjustments it is possible to provide, given a decrease in fuel consumption, a slower decrease in heat transfer (or even its preservation at the initial level), than a decrease in gas consumption, and accordingly a steeper decrease in the electric power while maintaining the total energy output to gas consumption ratio. All of the above has been calculated and stated without taking into account the substitution of gas for liquid fuels. In principle, such substitution, complete or partial, if it can be implemented, allows removing all obstacles to maintaining the required level of electricity and heat output under an emergency in the gas supply system.

3.3 The Importance of Uninterrupted Gas Supply to Thermal Power Plants During Emergencies in Gas Supply Systems

An estimate of social and economic damage due to failures to ensure normal gas supply to power plants of different groups (see item 3.1), or, essentially, due to failures to provide normal power and heat supply to consumers of the power plants under consideration, is of crucial importance for such assessment. In some of the published studies [27 and others] the authors state the problems of operation of industrial consumers and TPPs as cost-based when peak load controls are available. At the same time, various peak load control modes are used to level out the load profile, especially under conditions of interruptions in the supply of the required types of energy resources. With respect to social damage, of the utmost importance is, if there is no possibility to substitute gas for liquid fuels, to prevent complete termination of gas supply or a deep reduction in gas supply (by more than 40–45%) of combined heat and power plants (of the steam and gas turbines as well as combined cycle types), which are the only or the dominant (that is, providing at least 65–75% of the total amount) source of heat for heating residential buildings and priority social facilities (children's, medical, educational institutions, social service facilities for the disabled) of the respective settlement or an urban district. Essentially, this set of CHP plants includes all plants of Groups 14, 16, and 18, since practically all of them (except for a small number of purely industrial CHP plants) provide heat to the above consumers. Less deep (less than 40–45%) a reduction in gas supply to CHP plants of these groups is deemed acceptable (although, of course, it is undesirable), since in this case, firstly, production of electricity by CHP plants can be reduced being compensated by nuclear, hydraulic, and coal power plants, as well as gas-fired CHP plants temporarily converted to liquid fuel; secondly, heat supply can be restrained to a certain extent, including heating of industrial facilities and part of public buildings, as well as the load of ensuring domestic hot water supply. The top priority of gas consumers indicated above, that is heating and industrial-heating

CHP plants, is due to the risk of an unacceptably deep decrease in indoor temperature in the specified premises, the rate of cold-related diseases (sometimes followed by death), considerable deterioration of well-being of people.

The same priority is given to the provision of uninterrupted gas supply to CHP plants that in winter supply heat supply for air heating in the forced-air ventilation systems of chemical and other hazardous industries, coal pits, and mines. As the third component of the highest priority, one should list uninterrupted gas supply to power plants, that are the only or the dominant (in case of a weak electrical link with the electric power system) source of power supply to one or more settlements or essential production facilities.

It is also very important to keep at least a partial gas supply to CHP plants burdened with the task of heat supply to the settlement adjacent to the plant (Groups 4, 8, and 12), in the absence of a backup fuel oil (or other liquid fuel) system.

The next rank of priority should be assigned to large gas-fired gigawatt grade CPPs (i.e., with the capacity of more than 1 million kW or more than 1 GW) from groups 3, 4, 11, and 12, bearing in mind their role in ensuring the balance of the UES of Russia, a high probability of electricity (capacity) shortage on the national scale (or on the scale of a large region such as a federal district) when such a power plant is shut down due to the termination of gas supply to it, in the absence of a backup liquid fuel system.

Next, in terms of the importance of uninterrupted gas supply, come all other gas-fired TPPs that do not have a backup fuel oil (or other liquid fuel) system.

Gas-fired TPPs with a backup (emergency) fuel oil or other liquid fuel systems (they prevail in the mix of gas-fired TPPs in Russia) should be regarded as less critical with respect to ensuring their uninterrupted gas supply. However, even they should be ordered in descending order with respect to their priority in a way similar to that described above for TPPs with no backup liquid fuel system. The importance of setting priorities within this set of TPPs (Groups 1, 2, 5, 6, 9, 10, 13, 15, and 17) is due to, in addition to the above considerations, the risk of failures and accidents during gas substitution for liquid fuels.

3.4 Algorithms of the Rational Transition of Gas-Fired Power Plants from the Normal to a Non-standard Mode of Operation and Back (Restoring the Normal Mode)

The normal mode of operation of a gas-fired TPPs of any kind is understood to be the mode of its operation with running on the standard fuel, that is (pipeline) natural gas with the load profile corresponding to the dispatch schedule of generation (supply) of electricity and heat, including the maximum load mode corresponding to the rated electric and thermal capacity of the power unit (unit, power installation). A non-standard mode is a mode caused by a disruption of regular gas supply due to an emergency in the gas supply system and when the power plant is operating either on a backup liquid fuel (with the above-mentioned load according to the dispatch schedule) or on the standard fuel, that is gas, but with a reduced load, or with a power plant shutdown. The latter two options are implemented in case the power plant does not have a backup fuel (fuel oil, diesel fuel, etc.) system. Descriptions of the rational schedules considered below are presented by the homogeneous groups of TPPs, that is gas consumers, as defined in Sect. 3.1.

Group 1. Steam turbine gas-fired CPPs (modules, power units)[1] with a fuel oil system, not burdened with the secondary task (providing heat supply to the settlement adjacent to the plant)[1]. As mentioned above, the switchover of the power unit from gas to fuel oil is carried out within 40–60 min, when the fuel oil system is prepared in advance and when qualified operators and required instrumentation are available. Thus, the hourly transition schedule can be presented as a three-stage one: Stage I—standard gas consumption; Stage II—a gradual reduction in gas consumption to zero with a gradual increase in fuel oil supply up to the required level (up to 100%); Stage III—steady-state (if temporary) operation of the unit running on fuel oil. The hourly schedule of returning to the regular gas supply mode mirrors the above path (operating on fuel oil—a gradual change of the fuel mix with substitution of fuel oil for gas within 40–60 min—operating on gas).

Group 2. The same plants (units) as in Group 1 but burdened with the secondary task. Transition algorithms are absolutely the same as in Group 1 (Group 2, as well as some other groups of CHP plants, are formed because the power plants included therein rank somewhat higher than the plants that make up Group 1 in terms of the importance of their uninterrupted gas supply).

Group 3. Steam turbine CPPs, the same as in Group 1, but without a fuel oil system and not burdened with the secondary task. In case of a gas shortage due to an emergency in the gas supply system, such plants (units) are subject to either shutting down or (if they play a crucial role in ensuring the balance of electric power) they are switched to operation with reduced capacity and correspondingly reduced gas consumption. Both such variants of the transition to a non-standard mode are completed within a matter of an hour, as well as the reverse transition to a normal mode.

Group 4. Plants (units) are the same as in Group 3 but burdened with the secondary task. Unlike with Group 3, the option of shutting them down is not acceptable for the modules (units) of this group. Due to the presence of the secondary task, that is heat supply to the power plant settlement, they can only be unloaded (if necessary, as deep as possible, but with the preservation of the possibility of steam extraction from the regenerative extractions for heating of the settlement adjacent to the plant).

Group 5. Gas turbine stations (installations, units) of the "condensation" type with a backup liquid fuel system, not burdened with the secondary task. The switchover to a backup (emergency) fuel, that is diesel or other distillate fuel, is carried out faster than in the case of the steam turbine and takes 25–35 min. The switchover algorithm is as straightforward as it is for Group 1. The same applies to the algorithm of the reverse switchover, that is restoration of regular gas supply.

[1]These clarifications (in parentheses) will not be repeated further.

Group 6. The same GTUs as in Group 5 but burdened with the secondary task. The switchover algorithms are absolutely the same as in Group 5.

Group 7. GTUs of the "condensing" type, having no liquid fuel system, not burdened with the secondary task. In case of an emergency in the gas supply system, these units can be shut down, with subsequent start-up after the restoration of normal gas supply. Both of these procedures are quite fast to implement, the algorithms of the transition from one mode to another are straightforward.

Group 8. The same GTUs but burdened with the secondary task. As their non-standard mode, one should adopt the minimum capacity operating mode. The transition to such a mode and return to the regular mode are quick, the algorithms are rudimentary.

Group 9. CCGT-CPPs with a backup liquid fuel system, not burdened with the secondary task. Duration of the procedure of the switchover of one unit from gas to liquid fuel or that of the reverse switchover is 35–50 min The three-stage transition schedule described above for Group 1 is implemented here, as well as the same schedule of the reverse transition (restoring the normal mode).

Group 10. The same CCGT-CPPs but burdened with the secondary task. The transition algorithms are absolutely the same as in Group 9.

Group 11. CCGT-CPPs with no backup liquid fuel system, not burdened with the secondary task. All the considerations concerning the algorithms of the transition from the standard mode to a non-standard mode and back that are formulated for Group 3 are applicable here.

Group 12. The same CCGT-CPPs but burdened with the secondary task. All the considerations concerning the algorithms as formulated for Group 4 are applicable here.

Group 13. Steam turbine CHP plants (detailed in Sect. 3.1) with a backup fuel oil system. The non-standard mode for them is operating on fuel oil. The algorithms of the switchover from gas to fuel oil and the reverse transition are essentially the same as those considered for Group 1, but due to the lower unit capacity of CHP units (25–250 MW vs. 100–800 MW for CPPs), the switchover duration is somewhat shorter: 30–50 min versus 40–60 min for CHP plants.

Group 14. The same CHP plants as in Group 13 but having no backup fuel oil system. For a major part of them, a non-standard mode (in case of an emergency in the gas supply system) should be operating on gas with a reduced thermal load and the electrical power reduced as much as possible. The need to maintain gas supply to these plants is determined by their highest priority among gas-fired power plants, which in turn is due to unacceptable social damage from a deep reduction in the ambient temperature in residential units and premises of priority social facilities, as well as from disruptions in the operation of ventilation systems of hazardous (or dangerous) industries. At the same time, it is possible to reduce the heat load (and the corresponding gas consumption), if necessary, by means of the following: a significant reduction in heat consumption (or, in some case, its complete cessation) for the process needs of the industry; for heating and ventilation of non-priority industrial and public buildings; for domestic hot water supply, as well as due to some admissible reduction of the ambient temperature in the above top priority premises. In total, depending on the mix of heat consumers of a particular CHP, such a forced reduction in heat supply and the corresponding reduction in gas consumption may amount to 35–55% (i.e., up to 45–65% of the regular, or rated, consumption). The energy capacity of the CHP plant (electricity + heat) can be reduced to the minimum safe output of 40–50% of the rated capacity of boiler units. Thus, the gas consumption of the steam turbine CHP plants (one should be reminded that we are dealing here with CHP plants *without* a fuel oil system) should be kept at the level of 45–60% of the rated (standard) level. Such unloading, provided it is prepared in advance, may take up to 30 min. These figures determine the parameters of the transition from the normal to a non-standard gas supply mode and the corresponding algorithm. The algorithm of the reverse transition, that is restoration of the regular gas supply mode of the power plant, is similar.

As noted, all this applies to a major part of the CHP plants of Group 14. A small fraction of plants of this group, that is purely industrial CHP plants that do not have high priority heat consumers, can be unloaded more deeply (including shutting down of some of the units) or, if necessary, even shut down completely (taken out of service), which does not change, in general, the transition algorithms (from the normal to a non-standard mode and back) as outlined above for the main part of CHP plants of Group 14.

Group 15. Gas turbine CHP plants with a backup liquid fuel system. A non-standard mode: operating on a liquid fuel. The algorithms for the transition from the normal to a non-standard mode and back are the same as those specified for Group 5 (or 6).

Group 16. GTU-CHP plants with no backup liquid fuel economy. The non-standard mode for them is operating on gas with reduced electricity and heat load. Potential of a heat load reduction is the same as for steam CHP plants (Group 14): up to 45–65% of the standard load. In view of the more rigid link between electric and thermal capacity, the limit of a reduction in electric load and a reduction in gas consumption per unit is approximately the same, taking into account, on the one hand, the allowable limit of heat supply to certain categories of consumers; on the other hand, the unacceptability of a GTU capacity reduction below the minimum safe output (and, of course, assuming that other power plants, coal-fired, nuclear, hydraulic, as well as gas-fired CPPs and CHP plants temporarily switched over to liquid fuel, will ensure the electric load curve of the EPS). The duration of the transition to the non-standard mode is 15–25 min. The same applies to the reverse transition after the emergency is terminated.

Group 17. CCGT-CHP plants with a backup liquid fuel system. The non-standard mode for them is operating on liquid fuel. The algorithms of the transition from the normal mode to the non-standard mode and the reverse transition are the same as for Group 9.

Group 18. GTU-CHP plants with no backup liquid fuel system. The non-standard mode for them is operating on gas with electricity and heat load reduced to 45–65%. The algorithms of the transition from the normal to the non-standard mode and the reverse transition are the same as for Group 16.

Chapter 4
Control of Operation Modes of Gas Consumers in the Public Utility Sector Under Conditions of a Reduction, Termination, and Restoration of Gas Supply

4.1 Homogeneous Gas Consumer Groups in the Public Utilities Sector

From the point of view of gas consumption (and, actually, energy consumption in general), residential and public buildings, as well as facilities that provide these buildings with heat supply services (except for power generation facilities), cold and hot water supply, water disposal, and solid domestic waste disposal are classified as the public utilities sector.

Furthermore, residential buildings include single and multi-unit residential buildings, dormitories, hotels, barracks, as well as buildings for custodial detention (forced labor camps, prisons, and pre-trial detention facilities). Public buildings include buildings (premises) of educational, medical, recreational, children's, cultural, entertainment, and other leisure facilities; establishments (enterprises) of retail and wholesale trade, public catering, consumer and social services, mass and specialized information, communications, defense and security; administrative buildings (premises of representative, executive, judicial authorities, local self-government of public, financial, and insurance organizations). Usually (and this study is no exception) public buildings also include a certain range of industrial buildings located within the limits of urban area (but not at industrial sites of enterprises): research, design and survey, design, geological, construction, repair, and logistics organizations.

In residential and public buildings, direct gas receivers are the following: cooking appliances; water heaters; local heat generators (from house stoves to individual boilers). Thus, residential and public buildings use gas as a final energy carrier.

In terms of gas supply, only boilers, the only or the main fuel of which is natural gas (hereinafter referred to as gas-fired boilers) are of interest. From the point of view of the issues covered by this study, it is important to know not only the possibilities of controlling the modes of operation of the boiler houses themselves but also the behavior of heat consumers (with respect to their requirements and their response to a reduction and termination of gas supply). The considered set of boiler houses

V. I. Rabchuk et al., *Control of Operation Modes of Gas Consumers in the Event of Gas Supply Disruptions*, https://doi.org/10.1007/978-3-030-59731-3_4

includes, firstly, all gas-fired boiler houses serving only residential and public buildings (utility boiler houses), regardless of their form of ownership and affiliation): settlement, district, city quarter, group (serving several buildings), house (basement, attached, roof) boiler houses; secondly, the so-called industrial-heating boiler houses with a significant share of heat, supplied by them for heating of residential and public buildings. In our attempts to select the facilities of the public utilities sector in the most comprehensible way possible, we assumed that the above substantial share should be not less than 30%. Industrial-heating boiler houses with the share in the annual heat supply of the public utilities sector facilities (residential and public buildings) less than 30%, in this case, are considered and treated as industrial facilities (or facilities of other production sectors: agriculture, transport, and construction).

Almost all gas consumption by the two consumer groups in the public utilities sector considered (residential and public buildings and boiler houses proper) is its use for heating purposes: 100% of gas for boiler houses, at least 50% of gas for residential and public buildings (local heat generators and water heaters: heat supply only; cooking appliances also contribute to internal heat generation within premises). To this end in what follows we will review the role of gas in Russia's heating supply, particularly for the public utilities sector as based on statistical data. Gas accounts for 68% of the total fuel consumption for heating purposes. In heat production, the share of boiler houses (district heating and distributed heating sectors) in all types of fuel together with autonomous sources (the above local heat generators and water heaters) is 63% against 32% contributed by power plants. The share of natural gas used for heat production by boiler houses, together with the gas directly used by households, amounted to about 31% of all gas consumed in Russia in 2010, while all industries consumed about 26%.

Taking into account technical features and operation and process specifics of different types of facilities, that is gas consumers in the public utilities sector, their social functions and the degree of responsibility for preventing unacceptable damage from disruptions of normal gas supply, we provide for the division of these facilities into homogeneous groups of gas consumers based on the following attributes:

- A facility that is a consumer of gas as its final energy carrier (residential and public buildings) and a facility that is a consumer of gas for its conversion into the final energy carrier (boiler houses);
- The first category of facilities mentioned above covers facilities that differ in the availability or unavailability of each of the three types of gas receivers (cooking appliances, water heaters, local heat generators);
- Within the same category of facilities, we distinguish between the facilities that allow and that do not allow gas supply disruptions (interruptions);
- For the second category of facilities (public utility boiler houses), we distinguish between the facilities that have and do not have a backup (emergency) liquid fuel (fuel oil);
- Among public utility boiler houses, there are basic boiler houses that have their own mix of heat consumers (providing them with all heat their need), and peak

boiler houses that supplement the respective CHP plants (heating up hot water supplied by the CHP plant);

- Among public utility boiler houses, there are also boiler houses that have attached to them or not attached to them the highest priority heat consumers; as evidenced by the study of actual circumstances, almost all utility boiler-houses serve the high priority heat consumers; however, boiler houses supplying steam (hot water) only to baths, laundries, and similar public service companies or only some categories of public buildings (e.g., clubs, cinemas, some administrative buildings) may be classified as less critical gas consumers. These boiler house categories in what follows will be referred to as priority and non-priority boiler houses, respectively.

Taking these criteria into account, the following homogeneous groups of facilities, i.e., consumers of pipeline natural gas, are distinguished in the "Public utilities sector".

1. Residential premises (buildings) equipped with district heating (coming from CHP plants, boiler houses, other sources), centralized hot water supply from the same sources, floor gas stoves (ovens) that are the only type of the gas receiver there.
2. The same residential premises as in Group 1, but those where instead of a centralized hot-water supply system there are gas-fired water heaters (flow-type calorifiers) that are gas receivers along with gas stoves.
3. The same residential premises as in Group 2, but those where instead of central heating there are local gas heat generators that are gas receivers along with gas stoves and water heaters.
4. Catering facilities (restaurants, cafes, canteens, etc.) that cook hot meals and use gas only for cooking. Gas consumers of this group are both independent organizations (outlets) and catering facilities in hospitals, sanatoriums, educational, children's, administrative, and similar institutions, as well as those "on-site", that is within industrial, transport, and other manufacturing facilities.
5. The same public catering facilities as in Group 4, but using gas not only for cooking, but also for hot water supply.
6. The same facilities as in Group 5 but using gas for local heating by individual heat generators as well.
7. Public buildings that use local gas-fired heat generators for heating (mainly in rural and urban-type settlements) and belong to priority facilities that do not allow gas supply disruptions (interruptions or deep reductions).
8. The same buildings as in Group 7, but not belonging to the priority facilities and allowing for gas supply disruptions.
9. The same buildings as in Group 7 but using gas not only for heating but also for hot water supply.
10. The same buildings as in Group 8 but using gas not only for heating but also for hot water supply.

    Note. For all facilities of Groups 1–10 it is assumed, considering the actual state and designs of such facilities, that there is no possibility to substitute gas for liquid fuel and even less so for solid fuel.

11. Residential and public buildings intended for seasonal (summer) stay or use (summer cottages, cottages in garden and vegetable lot settlements, children's summer camps, etc.) that use gas for heating, cooking, and hot water supply.
12. Basic priority utility boiler houses equipped with a backup fuel oil (or other liquid fuel) system.
13. The same boiler houses but having no liquid fuel system.
14. Basic non-priority utility boiler houses equipped with a backup liquid fuel system.
15. The same boiler houses but having no liquid fuel system.
16. Peak utility boiler houses equipped with a backup liquid fuel system.
17. The same boiler houses but having no liquid fuel system.
    Note. In contrast to gas-fired CHP plants that are for the most part equipped with a fuel oil system, the actual stock of boiler houses in Russia consists of boiler houses, only a smaller part of which (in terms of their number and even total capacity) has a backup fuel oil system: these are some of the district gas-fired boiler houses and most of the industrial gas-fired boilers (with some of them being even gas- and fuel oil-fired). Small boiler houses, that is to say almost all of them, do not have a backup liquid fuel system; it is only seldom available in the case of medium-capacity boiler houses.

4.2 Special Aspects of Control of Operation Modes of Public Utility Sector Facilities Under Conditions of a Reduction, Termination, and Restoration of Gas Supply

Elements of facilities of the public utilities sector whose operation is impaired when the gas supply is reduced or terminated. Gas consuming facilities of the public utilities sector are, firstly, residential and public buildings (see Sect. 4.1 for the entities that make it up), where natural gas is the only energy carrier for cooking and (or) hot water supply (water heating) and (or) heating; secondly, utility boiler houses (in the sense of classifying boiler houses as public utilities explicated in Sect. 4.1), the main or the only fuel for which is pipeline natural gas.

Similar questions have been raised with regard to petrol stations. For example, in [28] and other studies. In general, for this group of consumers (by that we mean public utilities consumption) the issues of equalization of natural gas consumption mode by the so-called peak-shaving method have been the subject of active discussion [29]. In particular, it is proposed to achieve this by regulating gas prices by time zones in order to reduce the values of its peak consumption. The findings of the research carried out by the authors attest to the fact that maximum peak consumption after shaving the peaks by shifting the load can be reduced by 10–12% on average. These measures should also lead to some increase in the reliability of gas supply to consumers under conditions of high energy consumption.

For residential and public buildings (Groups 1–11), the elements directly responding to changes in gas supply are the following: cooking appliances (gas stoves, gas ovens, broilers); water heaters (including flow-type calorifiers); local heat generators (heating boilers, gas fireplaces, household stoves, and other space heaters). A while back, some of these issues were treated in detail in a number of studies. For example, [30] addressed the issue of monitoring of energy consumption in buildings in order to conserve energy resources, especially when supply is limited. For this purpose, a real-time energy management system was proposed.

For boiler houses (Groups 12–17) such elements are water heating and steam boilers, that is the main equipment of boiler houses. Since, according to Sect. 4.1, one of the types of boiler houses is referred to as the household boiler house, it is necessary to define more clearly the boundary between the latter and local heat generators (LHG): the LHG provides heat to a single-family dwelling house or one or more rooms in such a house, or a separate apartment in an apartment building, or sometimes a two-family dwelling house, as well as a separate establishment or separate public premises. Heat sources of a larger scale belong to boiler houses. Another distinguishing feature is the entity that performs maintenance of the heat source: the LHG is serviced by the owner (user) of the premises, the boiler house is serviced by specialized personnel.

For boilers of any type and any group of Groups 12–17 it is possible (technically permissible) to gradually decrease gas supply within adjustment range limits of the boiler unit down to its safe minimum output, that is usually down to 50% of the rated gas consumption and the corresponding decrease in heat output. A deeper reduction in gas consumption and all the more so sudden complete halt of gas supply to the boiler unit is unacceptable: such actions lead to an emergency. In order to avoid this (if no substitution of gas for a backup liquid fuel is provided for) during the planned shutdown of the boiler (either some of the boilers, or the boiler house as a whole) due to a reduction or termination of gas supply, it is necessary, after the gradual unloading of the minimum safe output, to maintain it in operation for a short time (up to 1 h) with the subsequent shutdown in accordance with the instruction (procedure), usually during 15–20 min.

In contrast to the above (the case of boilers), gas receivers in buildings (cooking appliances, water heaters, and local heat generators) (consumers of Groups 1–11) can technically be shut down (almost instantaneously) without emergencies (although shutdown of consumers of Groups 1–10 is usually inadmissible for other reasons: unacceptable social damage—see Sect. 4.3; it seems acceptable to disconnect consumers of Group 11 from gas supply in case of an emergency in the gas supply system).

The switchover of gas consumers in the public utilities sector from gas to other energy sources and the reverse switchover: its possibility and duration. In Sect. 4.1 it was noted that switching over of gas receivers in residential and public buildings to other fuels is not provided for by design. However, in certain cases, some gas consumers are ready on their own, if the power supply system (including domestic one) allows that, to switch over to electrical appliances for cooking and water heating, sometimes even to electric heating. However, such extraordinary situations should not be considered in this case. Thus, in relation to this set of facilities of the public utilities sector (Groups 1–11) we assume that their switchover from gas to other fuels or other energy carriers is impossible.

The situation is different in the case of utility boiler houses (gas consumers of Groups 12–17). Some district (settlement) and industrial heating gas boilers are equipped with boilers with gas and fuel oil burners and a backup (emergency) fuel oil system (in some cases it is possible to burn distillate fuel, e.g., diesel fuel) (the boilers here are gas consumers of Groups 12, 14, and 16). Due to much lower unit capacity of the boilers of the mentioned boiler houses as compared to CHP plants (and even more so CPPs), simpler thermal and process schemes of the boiler houses, their lower parameter values, the duration of the switchover from gas to liquid fuel of an individual boiler is somewhat shorter than indicated in Sect. 3.2 for power plants, or rather an individual power unit or module (40–60 min). For the boiler unit of the considered boiler houses (Groups 12, 14, and 16) the procedure of its switchover from gas to fuel oil, depending on the boiler capacity, its condition, the initial condition of the fuel oil system, qualification of personnel, and other circumstances may take from 15 to 30 min. Approximately the same is the duration of the reverse transition from fuel oil to gas after the restoration of normal (regular) gas supply. In case of necessity of a one-time switchover of all boiler units of a multi-unit boiler house from gas to fuel oil (or from fuel oil to gas), the duration of such switchover is 30–60 min. Boiler houses of Groups 13, 15, and 17 that lack a backup liquid fuel system, and this includes almost all small boiler houses (house-, group-, and some of the city quarter boiler houses), a significant part of the medium capacity boiler houses, some of the larger boiler houses, under conditions of a termination or deep reduction in gas supply to the settlement as a result of an emergency in the gas supply system should either be kept in operation running on gas or stopped (see Sect. 4.4).

Dependence on the level of gas undersupply of gas consumers' output in the public utilities sector, that is supply of heat by utility boiler houses, provision of services of heating, hot water supply, heating up of food and cooking by local heat generators, water heaters, cooking appliances. Under emergency conditions, the gas supply system reduces, sometimes to zero, the gas supply to consumers of some (any) settlement of the corresponding federal subject or region of the country. As a result, gas consumers in the public utilities sector are switching over to a non-standard mode of operation. Some of them switch over to a backup liquid fuel, others can be stopped (disconnected), while still others should be kept in operation, but with a decrease in gas supply and a corresponding reduction in the output of products, that is the provision of relevant services.

Gas-fired utility boiler houses can be operated in an accident-free way within adjustment range limits of their boilers (steam or hot water ones), the lower limit of which, that is the minimum safe output, is, as noted, about 50% of the rated boiler output, oscillating around this value in the range of 40–55%, depending on the type of boilers, their unit capacity, and the features of their process scheme. It can be roughly assumed that one percent of undersupply corresponds to one percent of undersupply of heat. Approximately the same is the situation with local heat generators and individual water heaters, but the allowable reduction in their thermal capacity, as well as the capacity of utility boiler houses, is determined not by the specified technical possibilities, but by sanitary and hygienic requirements (standards), which allow not so deep (up to 50%) a reduction in heat supply (see Sect. 4.3, 50 for more details).

In conclusion, it should be noted that if there is a possibility to substitute gas for fuel oil (or other liquid fuel) at utility boiler houses, the undersupply of gas to them will certainly not affect the output of these boilers in any way.

4.3 Importance of Uninterrupted Gas Supply to the Public Utilities Sector (Boiler Houses, Residential and Public Buildings) During Emergencies in Gas Supply Systems

Just like in the case with power plants (Chap. 3), crucial for this assessment is the requirement to mitigate the social damage from disruptions of normal gas supply to public utility facilities or essentially from disruptions of normal heat supply to gas consumers, both boiler houses and residential and public buildings as such (assuming somewhat tentatively that the medium temperature processes of cooking are also the processes of heating). In terms of mitigation of social damage, prevention of unacceptable social damage, of utmost importance (top priority) is to prevent complete termination or a deep reduction in gas supply to the three consumer groups: (a) residential buildings (premises): facilities of Groups 1, 2, and 3 (see Sect. 4.1); (b) public buildings that are priority facilities that do not allow serious disruptions (a termination or deep reduction) of gas supply/heat supply (such facilities include childcare, medical, and educational facilities, as well as social service facilities for the disabled), that is facilities of Groups 7 and 9, as well as public catering facilities of Groups 4–6 located at these facilities; (c) boiler houses that serve as the only or dominant source of heat for heating and hot water supply of residential buildings and/or priority social facilities of the same type as those listed in item "b" that do not allow serious disruptions of heat supply, its termination or deep reduction of heat supply (facilities of Group 13 that as evidenced by their study, includes the majority of utility boiler houses that do not have a backup fuel oil system). The highest priority of these three sets of gas-consuming facilities is due to the risk of the following (as is the case with CHP plants—see Sect. 3.3): an unacceptably deep drop of indoor temperature in dwellings and premises of priority social facilities; an unacceptable deterioration in the comfort of staying in these premises (not only due to the problem of temperature conditions but also due to the issues related to cooking and hot water supply); a significant deterioration of people's well-being; outbreaks of cold-related diseases (sometimes followed by death).

At the same time, it should be noted that some boiler houses of Group 13 serve not only residential buildings and priority social facilities but also public buildings and production facilities of less priority that during an emergency in the gas supply system can either be closed or transferred to the mode of degraded heat supply and at this cost as well as that of the allowable (by 2–3 °C) decrease in temperature in these priority premises, gas supply to some of the boiler houses of Group 13 can be slightly reduced (although it is still undesirable).

Of the same priority as supplying gas to facilities of sets "a-b-c" is to ensure uninterrupted gas supply to boiler houses (those that lack a backup fuel system), providing heat supply for air heating in forced ventilation systems of chemical and other hazardous industries, coal pits, and mines.

The next, lower priority rank of uninterrupted gas supply to facilities of the public utilities sector (essentially, uninterrupted heat supply) is assigned, firstly, to public catering facilities (facilities of Groups 4–6) with the exception of that part of this set that is linked to the operation of priority social facilities and is designated above as belonging to set "b"; secondly, non-priority public buildings (facilities of Groups 8 and 10); thirdly, boiler houses (Group of facilities 15) that supply heat *only* to non-priority buildings, sometimes with the addition of production facilities as heat consumers. Let us notice the following: most utility boiler houses serve residential and/or priority social facilities, oftentimes along with non-priority public buildings and production facilities. Some of such non-priority gas-consuming facilities can be (during an emergency in the gas supply system) closed or stopped, but most of them, that are, after all, necessary for maintaining normal life activities, can be transferred only to a reduced gas consumption mode.

At the third stage in terms of the importance of uninterrupted gas supply, there should be basic utility boiler houses with a backup (emergency) fuel oil or some other liquid fuel system, as well as peak utility boiler houses (facilities of Groups 12, 14, 16, and 17). The same stage should include facilities for seasonal (summer) stay (use), that is facilities of Group 11. Priority ranking (related to preserving gas supply in a reduced volume or switching over the boiler house to fuel oil during the emergencies covered by this study) among the third stage facilities should be done first of all based on local conditions, however, factoring in as well the principles for gas-consuming facilities lacking a backup fuel system as formulated for Russia as a whole.

4.4 Algorithms of the Rational Transition of Gas Consumers of the Public Utilities Sector from the Regular Mode of Operation to a Non-standard Mode and Back (Restoring the Normal Mode)

The standard mode of operation of a gas-fired utility boiler house here is understood to be the mode of its operation with running on the standard fuel, which is pipeline natural gas, with the load adequate to meet fully the demand of consumers of the boiler house for heat for the purposes of heating, ventilation, air conditioning, hot water supply, while complying with standards on temperature, humidity, and air purity in the premises and taking into account weather conditions, including the mode of maximum load corresponding to the rated (design, calculated) thermal capacity of the boiler house (its boilers). The standard mode of gas supply of residential or public buildings here is understood as a mode that ensures meeting completely their needs in gas for cooking appliances, gas water heaters, gas-fired local heat generators, taking into account normative values of coefficients of simultaneity (non-uniformity) of gas consumption and taking into account, in relation to heat generators, requirements for temperature in premises, air exchange, conditions.

A non-standard mode of operation of a gas-fired public utility boiler house is understood here as a mode caused by a disruption of normal gas supply, which is an emergency in the gas supply system, with a reduction or termination of gas supply to a given locality, and consisting in the operation of a boiler house as running either on a backup (emergency) liquid fuel or gas but with reduced consumption of gas and, accordingly, a reduced heat load of the boiler house relative to its demand, or a shutdown of the boiler house or some of its boiler units. A non-standard gas supply mode of residential or public buildings (including public catering facilities) is understood here as the mode caused by the same factors as those listed above for the boiler house, and consisting either in the operation of cooking appliances, gas water heaters, and gas local heat generators with reduced gas consumption and, accordingly, their energy output and performance or in the temporary cessation of operation of these gas receivers. Descriptions of the rational schedules considered below are presented by the homogeneous groups of gas consumers of the public utilities sector, as defined in Sect. 4.1.

Group 1. Residential premises (buildings) with cooking appliances as the only gas consumer. No switchover to other energy carriers is possible. Switching off (gas supply termination) is not permitted. It is technically unsafe to reduce gas supply by reducing the pressure at the gas distribution station (GDS) or gas distribution point (GDP). Thus, in the event of an emergency in the gas supply system it is still necessary to preserve the supply to consumers of this group, and only in extreme cases (the need to disconnect the locality as a whole), with some restraint and with prior notification, to switch them off. This transition to a non-standard mode is almost instantaneous, as is restoring the normal mode.

Group 2. Residential premises that use gas not only by cooking appliances but also by water heaters. The situation is identical to that with Group 1 consumers.

Group 3. Residential premises that use gas for cooking appliances, water heaters, and local heat generators. The situation is similar to that of Group 1 consumers.

Group 4. Public catering facilities that use gas only for cooking. No switchover to other energy carriers is possible. The shutdown is not allowed, a reduction in gas supply is technically unacceptable. In case of an emergency, it is necessary to maintain gas supply to consumers of this group. Only as a last resort (see Group 1) is it possible to switch off this group of consumers (together with the entire locality or its major area).

Group 5. The same facilities that use gas for cooking and water heating. The situation is similar to that of Group 4.

Group 6. The same facilities that use gas for cooking, water heating, and local heating. The situation is similar to that of Group 4.

Group 7. Public buildings that are priority facilities equipped with gas-fired heat generators. The situation is similar to that of Group 1 consumers.

Group 8. Public buildings that are non-priority facilities equipped with gas heat generators. A significant share of such facilities can be shut down under conditions of an emergency in the gas supply system, but the list of buildings subject to shutting down and the shutdown procedure should be worked out and agreed upon in advance. Such a transition to a non-standard mode itself is carried out almost immediately and smoothly (as well as restoration of the normal mode), but it is desirable to avoid it because it causes certain (if acceptable) social and material damage.

Group 9. The same buildings as in Group 7 but using gas not only for heating but also for water heating. The situation is similar to that of Group 7 consumers.

Group 10. The same buildings as in Group 8 but using gas not only for heating but also for water heating. The situation (transition algorithms) is similar to that of Group 8 consumers.

Group 11. Residential and public buildings intended for seasonal (summer) residence (use), using gas for local heating, hot water supply, and cooking. In case of an emergency in the gas supply system, these facilities may be shut down. Such a transition (the probability of its occurrence is small, as major emergencies usually take place in winter) to a non-standard mode (by all means agreed upon and prepared in advance) is performed almost immediately and smoothly, as well as restoration of the normal mode.

Group 12. Basic priority utility boiler houses with a backup liquid fuel system. In case of an emergency in the gas supply system, the boiler house switches over to a backup fuel—this is the transition to a non-standard mode. Transition duration, as indicated above, is 15–30 min. The transition is performed smoothly. The same is the duration and quality of the reverse switchover from fuel oil to gas.

Group 13. The same boiler rooms but lacking a fuel oil system. Under emergency conditions, the boiler house should be kept in operation, with a reduction in heat output (to the safe minimum output of boilers at the most extreme case) due to a certain reduction in heat supply to non-priority public buildings and production facilities connected to it, as well as due to a 2–3 °C decrease in indoor temperature in residential and public buildings, and a corresponding decrease in gas consumption. In this case, the unloading of the boiler house is allowed within adjustment range limits of its boilers with no more than 50% on average. Gas consumption by the boiler house can be reduced within the same limits.

Group 14. Basic non-priority utility houses with a backup liquid fuel system. A non-standard mode is operating on liquid fuel (usually fuel oil). The duration and nature of the transition from the normal to the non-standard mode and back is similar to Group 12.

Group 15. The same boiler houses but lacking a backup fuel oil system. In case of an emergency, the boiler house can be either kept in operation with a reduction in heat output (and gas consumption) to the safe minimum output of boilers (see the description of Group 13) or stopped for the duration of the emergency (in case of an acute gas shortage). The duration of the shutdown, in accordance with the regulations of boiler houses is 15–20 min (after maintaining the operation at the safe minimum output level for about 60 min). The same is approximately the duration of the start-up of the boiler after the emergency, which is the restoration of the normal mode.

Group 16. Peak (contributing to water heating at the CHP) utility boiler houses with a backup liquid fuel system. Algorithms of the transition from the standard mode of fuel supply to a non-standard mode and back are similar to those of boiler houses of Group 14.

Group 17. Peak public utility boiler houses lacking a backup liquid fuel system. Algorithms of the transition from the standard mode of fuel supply to a non-standard mode and back are similar to those of boiler houses of Group 15.

Conclusion

1. This study deals with the issues of controlling the modes of operation of various natural gas consumers in the industrial, electric power, and public utilities sectors under conditions of a reduction, termination, and restoration of gas supply to them in case of large-scale disruptions in the operation of gas supply systems. As a result of the study, we have arrived at the data on *allowable restrictions* (up to the possibility of a complete shutdown) on *gas supply to such consumers* without initiating their emergency shutdown, i.e., in with the prevention of significant economic losses for the gas consumers themselves and for the country's economy. The obtained findings also relate to process conditions of the fail-safe shutdown for the gas consumers under consideration (required time for complete shutdown of the gas consumer and minimum level of gas supply to this consumer at the moment preceding the shutdown).
2. The above data (see Item 1) were used to substantiate the procedure (algorithm) of the rational dynamics of the transition of the gas consumer of each given type (from among the considered ones) from the normal mode of operation to the mode of operating under conditions of the emergency and to substantiate the algorithms of the rational dynamics of restoring the normal mode at the same consumers after the emergency is terminated. When substantiating the algorithms, the social and economic significance of the gas consumer was factored in as well. In this case, both the algorithm of the rational transition from the standard mode to a non-standard mode and the algorithm of restoration of the standard mode are hourly schedules of required volumes of gas supply to this gas consumer for the whole period of the transition from one mode of operation to another. In the first case (the transition from the normal to a non-standard mode), these are the minimum required volumes, while in the other case these are the volumes required to quickly restore the normal mode yet being subject to constraints associated with the operation and process specifics of the given gas consumer.
3. Due to the sheer diversity of gas consumers in the industrial sector, the study covered the most gas-intensive representatives of this sector:

V. I. Rabchuk et al., *Control of Operation Modes of Gas Consumers in the Event of Gas Supply Disruptions*, https://doi.org/10.1007/978-3-030-59731-3

   - key gas-consuming facilities of the integrated steel mill;
   - gas-consuming facilities of the non-ferrous smelting industry, such as aluminum, copper, lead, zinc, and nickel smelting, titanium and magnesium industry enterprises;
   - cement production based on the use of natural gas;
   - production of methanol and ammonia: representative facilities of gas consumers in petrochemical and agrochemical industries that make use of natural gas as a feedstock.

4. The main findings of the research aimed at solving the problems of control of the operation modes of gas consumers in the industrial sector are as follows:

   - rational schedules (algorithms) of the transition from the normal mode of operation to a non-standard mode during an emergency and the reverse transition to the normal mode after the emergency is terminated as applied to the following:

     - the integrated steel mill;
     - some of the production facilities of the iron and steel industry (sinter plants, coke production, rolling mills, steelmaking) in cases where they operate in a stand-alone way (outside of the integrated plant) and where they operate as elements of the integrated mill;
     - aluminum production (including such gas consumers as sintering and calcination furnaces for alumina production, fluoride salt furnaces, and electrode production shop);
     - copper smelters.

   - It was shown that in order to ensure the normal operation of the integrated steel mill even during the emergency period (albeit with some deterioration in economic performance) the mill must be supplied during the emergency with natural gas in the amount of not less than 30% of its average hourly supply under normal conditions; this 30% is consumed by steelmaking, where (unlike with other gas-consuming facilities of the iron and steel industry) substitution of natural gas for other fuel is impossible;
   - It has been shown that without significant economic damage it is possible to shut down certain iron and steel industry production facilities in a fail-safe way, if the latter (these production facilities) operate in a stand-alone mode, outside of the integrated mill); moreover, steelmaking requires 11–12 h for its shutdown with a reduction in gas supply from 100 to 50%, followed by a complete termination of this supply; coke production requires up to 16 h for its shutdown, and rolling mills take up to 2 h;
   - Conditions (required time, levels of gas supply) of the transition from the normal mode to a non-standard mode and back to the normal mode (after the emergency is terminated), conditions of the complete fail-safe shutdown (during an emergency in the NGS), as well as conditions and possibilities of continuation of operation even during an emergency in the NGS of the following types of enterprises:

- lead, zinc, nickel, and titanium and magnesium smelters;
- cement production;
- co-producing of methanol and ammonia, based on the methane conversion process in the presence of oxygen (where natural gas is used as a feedstock).

5. Gas consumers in the electric power industry and public utilities sector (including the direct use of gas by households) are *covered in the present study exhaustively* because today, taken together, they consume about 80% of all gas used in the country, and secondly, the number of their homogeneous groups is not as large as in the industrial sector. (A homogeneous group of gas consumers in our case is gas consumers that share the type of products or services they provide, operation and process specifics, and social and economic significance.)
6. For the purposes of studying special aspects of the behavior of gas consumers in the electric power industry under conditions of a reduction, termination, and restoration of natural gas supply to them, all these gas consumers (TPPs that serve different purposes and adopt different processes) were divided into 18 homogeneous groups. Considering the operation and process specifics of various TPPs, as well as the social and economic importance of their uninterrupted operation during an emergency, the following was determined for each homogeneous group of TPPs:

    - conditions of their complete fail-safe shutdown (time required for the shutdown and minimum levels of gas supply to the TPP at the moment immediately preceding the shutdown);
    - conditions of continuing operation of gas-fired TPPs even during an emergency for gas-fired TPPs lacking an auxiliary fuel system (minimum required levels of gas supply to them during an emergency);
    - duration of the switchover of gas-fired TPPs during an emergency to a backup fuel (for TPPs with an auxiliary fuel system);
    - the procedure (algorithm) of the transition to a non-standard mode and the algorithm of restoring the normal mode of TPP operation after an emergency in the gas supply system is terminated.

7. All gas consumers of the public utilities sector (various gas-fired boiler houses and direct use of gas by households) were divided into 17 homogeneous groups so as to facilitate the analysis of their behavior under conditions of a reduction, termination, and restoration of gas supply to them. For each such group, taking into account the requirements on the levels and degree of the importance of uninterrupted heat supply to its consumers (including hot water supply) by gas-fired boiler houses, as well as taking into account the social importance of uninterrupted gas supply to households directly (including individual heating systems and individual hot water installations), the following has been determined:

    - the list of homogeneous groups in the public utilities sector that can be disconnected from gas supply during an emergency in gas supply systems without any noticeable consequences;

- conditions for continuing operation of certain groups of gas consumers in the public utilities sector even during emergencies;
- duration of the switchover of gas consumers in the public utilities sector to a backup energy carrier (if there is such a possibility of the switchover);
- the procedure (algorithms) of the rational transition of different gas consumers in the public utilities sector from the normal mode of their gas supply to a non-standard mode and restoration of the normal mode.

References

1. W. Yu, J. Gong, Sh. Song, W. Huang, Y. Li, J. Zhang, B. Hong, Y. Zhang, K. Wen, X. Duan, Gas supply reliability analysis of a natural gas pipeline system considering the effects of underground gas storages. Appl. Energy **252**, 113418 (2019). https://doi.org/10.1016/j.apenergy.2019.113418
2. H. Su, E. Zio, J. Zhang, L. Chi, X. Li, Z. Zhan, A systematic data-driven demand side management method for smart natural gas supply systems. Energy Convers. Manag. **185**, 368–383 (2019). https://doi.org/10.1016/j.enconman.2019.01.114
3. C.A. Saldarriaga, H. Salazar, Security of the Colombian energy supply: the need for liquefied natural gas regasification terminals for power and natural gas sectors. Energy **100**, 349–362 (2016). https://doi.org/10.1016/j.energy.2016.01.064
4. Zh Qiao, Q. Guo, H. Sun, Optimal gas storage capacity in gas power plants considering electricity and natural gas systems constraints. Energy Procedia **142**, 2983–2989 (2017). https://doi.org/10.1016/j.egypro.2017.12.367
5. S. Kan, B. Chen, J. Meng, G. Chen, An extended overview of natural gas use embodied in world economy and supply chains: policy implications from a time series analysis. Energy Policy **137**, 111068 (2020). https://doi.org/10.1016/j.enpol.2019.111068
6. A. Prahl, K. Weingartner, Chapter 3—A study of Russia as key natural gas supplier to Europe in terms of security of supply and market power. Low-carbon Energy Secur. Eur. Perspect. 43–79 (2016). https://doi.org/10.1016/B978-0-12-802970-1.00003-6
7. B. Lin, Y. Kuang, Natural gas subsidies in the industrial sector in China: national and regional perspectives. Appl. Energy **260**, 114329 (2020). https://doi.org/10.1016/j.apenergy.2019.114329
8. S. Senderov, A. Edelev, Formation of a list of critical facilities in the gas transportation system of Russia in terms of energy security. Energy (2019). https://doi.org/10.1016/j.energy.2017.11.063
9. N.I. Voropai, S.M. Senderov, A.V. Edelev, Detection of "bottlenecks" and ways to overcome emergency situations in gas transportation networks on the example of the European gas pipeline network. Energy (2012). https://doi.org/10.1016/j.energy.2011.07.038
10. J.P. Deane, M.Ó. Ciaráin, B.P.Ó. Gallachóir, An integrated gas and electricity model of the EU energy system to examine supply interruptions. Appl. Energy **193**, 479–490 (2017). https://doi.org/10.1016/j.apenergy.2017.02.039
11. P. Jirutitijaroen, S. Kim, O. Kittithreerapronchai, J. Prina, An optimization model for natural gas supply portfolios of a power generation company. Appl. Energy **107**, 1–9 (2013). https://doi.org/10.1016/j.apenergy.2013.02.020

V. I. Rabchuk et al., *Control of Operation Modes of Gas Consumers in the Event of Gas Supply Disruptions*, https://doi.org/10.1007/978-3-030-59731-3

12. Sh.Kh. Basiri, F.M. Sobhani, S.J. Sadjadi, Developing natural-gas-supply security to mitigate distribution disruptions: a case study of the National Iranian Gas Company. J. Clean. Prod. **254**, 120066 (2020). https://doi.org/10.1016/j.jclepro.2020.120066
13. L. Shirazi, M. Sarmad, R.M. Rostami, P. Moein, M. Zare, Kh. Mohammadbeigy, Feasibility study of the small scale LNG plant infrastructure for gas supply in north of Iran (Case Study). Sustain. Energy Technol. Assess. **35**, 220–229 (2019). https://doi.org/10.1016/j.seta.2019.07.010
14. W. Ahmad, J. Rezaei, L.A. Tavasszy, M.P. de Brito, Commitment to and preparedness for sustainable supply chain management in the oil and gas industry. J. Environ. Manag. **180**, 202–213 (2016). https://doi.org/10.1016/j.jenvman.2016.04.056
15. W. Yu, Sh Song, Y. Li, Yu. Min, W. Huang, K. Wen, J. Gong, Gas supply reliability assessment of natural gas transmission pipeline systems. Energy **162**, 853–870 (2018). https://doi.org/10.1016/j.energy.2018.08.039
16. V.A. Krivandin, V.A. Artykhov, B.S. Mastrykov, et al., *Metallurgical Thermal Engineering*, vol. 1. Theoretical foundations: Undergraduate-level textbook (Metallurgiya, Moscow, 1986), 424 pp. (in Russian)
17. V.A. Krivandin, I.N. Nevedomskaya, V.V. Kobakhisze, et al., *Metallurgical Thermal Engineering*, vol. 2. Design and operation of furnaces: Undergraduate-level textbook (Metallurgiya, Moscow, 1986), 592 pp. (in Russian)
18. L. Teplov, Metal Forum 2007. Metallurgichesky Kompas (10), 5–9 (2007) (in Russian)
19. O.N. Bagrov, B.M. Kleshko, V.V. Mikhailov, *Energy Systems of Key Production Facilities of the Non-ferrous Smelting Industry* (Metallurgiya, Moscow, 1979), 376 pp. (in Russian)
20. O.N. Bagrov, V.P. Andreyev, V.I. Deyev, I.M. Rushchuk, *Energy Use in the Non-ferrous Smelting Industry* (Metallurgiya, Moscow, 1990), 112 pp. (in Russian)
21. V. Pereushanu, M. Korobya, G. Muska. *Production and Use of Hydrocarbons* (Khimiya, Moscow, 1987), 288 pp. (in Russian)
22. A.A. Ionin, *Gas Supply Systems: Undergraduate-Level Textbook* (Stroyizdat, 1989), 439 pp. (in Russian)
23. N.L. Staskevich, G.N. Severinets, Ya.Ya. Vigdorchik, *Handbook on Gas Supply and Gas Use* (Nedra, Leningrad, 1981), 762 pp. (In Russian)
24. *The Energy Industry of Russia: A Glimpse into the Future* (Supporting materials for the Energy Strategy of Russia to 2030) (Energiya Publishing House, Moscow, 2010), 616 pp. (in Russian)
25. *Operation and Development of the Electric Power Industry of the Russian Federation in 2010.* Information and analytical report of the Russian Ministry of Energy of the Russian Federation. Appendices (The edition of the Energy Forecasting Agency, Moscow, 2011), 265 pp. (in Russian)
26. N. Iliopoulos, M. Esteban, Sh. Kudo, Assessing the willingness of residential electricity consumers to adopt demand side management and distributed energy resources: a case study on the Japanese market. Energy Policy **137**, 111169 (2020). https://doi.org/10.1016/j.enpol.2019.111169
27. H. Khodaei, M. Hajiali, A. Darvishan, M. Sepehr, N. Ghadimi, Fuzzy-based heat and power hub models for cost-emission operation of an industrial consumer using compromise programming. Appl. Thermal Eng. **137**, 395–405 (2018). https://doi.org/10.1016/j.applthermaleng.2018.04.008
28. Y. Liu, Zh. Kong, Q. Zhang, Failure modes and effects analysis (FMEA) for the security of the supply chain system of the gas station in China. Ecotoxicol. Environ. Saf. **164**, 325–330 (2018). https://doi.org/10.1016/j.ecoenv.2018.08.028
29. L. Li, Ch. Gong, Sh. Tian, J. Jiao, The peak-shaving efficiency analysis of natural gas time-of-use pricing for residential consumers: evidence from multi-agent simulation. Energy **96**, 48–58 (2016). https://doi.org/10.1016/j.energy.2015.12.042
30. N. Haidar, M. Attia, S.-M. Senouci, El.-H. Aglzim, A. Kribeche, Z.B. Asus, New consumer-dependent energy management system to reduce cost and carbon impact in smart buildings. Sustain. Cities Soc. **39**, 740–750 (2018). https://doi.org/10.1016/j.scs.2017.11.033

Printed in the United States
by Baker & Taylor Publisher Services